From El Billar to Operations Fenix and Jaque:

The Colombian Security Force Experience, 1998–2008

Robert D. Ramsey III

Occasional Paper 34

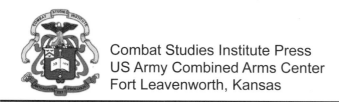

Combat Studies Institute Press
US Army Combined Arms Center
Fort Leavenworth, Kansas

Library of Congress Cataloging-in-Publication Data

Ramsey, Robert D., 1946-
    From El Billar to Operations Fenix and Jaque : the Colombian security force experience, 1998-2008 / by Robert D. Ramsey III.
        p. cm. -- (Occasional paper ; 34)
    Includes bibliographical references.
    1. National security--Colombia. 2. Internal security--Colombia. 3. Insurgency--Colombia. 4. Colombia--Armed Forces. 5. Colombia. Ejército. 6. Military assistance, American--Colombia. 7. Pastrana Arango, Andrés. 8. Uribe Vélez, Álvaro, 1952- 9. Colombia--Military policy. 10. Colombia--Politics and government--1974- I. Title. II. Series.

    UA625.R36 2009
    986.106'35--dc22

                                                                2009032483

First printing: December 2009

Foreword

Recent operations in Iraq and Afghanistan have given the US military an appreciation of both the importance and the challenges of working with and through host nation security forces in the aftermath of major combat operations. Secretary of Defense Robert M. Gates has indicated that these types of efforts will be an ongoing military requirement for the foreseeable future. The US military effort in support of Colombian security forces offers a different and a lesser known experience from which to learn: one that has been long-term, low-key, and seemingly successful. Between 1998 and 2008, Colombian security forces dramatically improved as they moved from what many considered the brink of disaster to being on the verge of victory.

The Combat Studies Institute (CSI) is pleased to publish its 34th Occasional Paper, *From El Billar to Operations Fenix and Jaque: The Colombian Security Force Experience, 1998–2008*. Written at the request of US Southern Command, this study begins with an overview of the general security situation prior to 1998, then traces Colombian and US efforts during the Pastrana presidency and Plan Colombia, and concludes with the subsequent actions of the Uribe administration. In the final section, the author offers observations from the Colombian experience for those in the US military who will be called on to work with and through host nation security forces in the future.

Although the Colombian experience offers no simple model to be replicated mindlessly elsewhere, it does offer observations and analysis that may be useful to the military professional struggling to address similar situations. Robert Ramsey's Occasional Paper not only provides a useful reminder of the inherent challenges for both host nation and US military personnel in those situations, but also shows that, given the right people, the right programs, and sufficient time, those challenges can be met successfully. Perhaps that is the best insight of all to be drawn from a reading of this work. *CSI—The Past Is Prologue!*

Douglas M. Fraser
General, US Air Force
Commanding General
US Southern Command

Acknowledgments

No one completes a study such as this by oneself. Among those deserving special thanks are the Combat Studies Institute Director Dr. William G. Robertson for assigning me this project; the United States (US) Southern Command historian Dr. Bradley Coleman for his assistance in gaining access to Colombian and United States personnel; the US Special Operations Command South historian Mr. Edward J. Dillenschneider for his assessment and for providing me interviews with personnel with Colombian experience; the US Army South G2 Colonel Mark A. Costello for arranging meetings with personnel with Colombian service; the Commandant of the Western Hemisphere Institute for Security Cooperation Colonel Felix L. Santiago and his senior Colombian officer, Lieutenant Colonel Jorge I. Monsalve, for the opportunity to interview Colombian students; the US Military Group-Colombia commander Colonel Kevin D. Saderup and Lieutenant Colonel Barbara R. Fick and US Navy Lieutenant Harry Watkins for scheduling and supporting meetings with senior Colombian civilian, military, and police officials; the Colombian Ministry of Defense for permitting research in Colombia from 1 to 7 February 2009 and attendance at its "Contemporary Counter-Terrorism and Counter-Insurgency Conference: the Colombian Experience" in Bogotá 29 March to 3 April 2009; the first two Colombian Army officers attending the School for Advance Military Studies, Lieutenant Colonel Juan M. Padilla and Lieutenant Colonel Juan C. Correa, for responding to my questions; the Combined Arms Research Library archivist John Dubuisson for his assistance in gathering materials; Robin Kern for the graphics; and Betty Weigand for her careful editing that improved the manuscript. Without the support and effort of those above, as well as many others, I could not have completed this work in the time permitted. Responsibility for errors in fact or judgment rests with me alone.

Contents

Tables

Figures

Chapter 1

The Colombian Security Situation

This [defeat at El Billar in March 1998] is without a doubt the biggest defeat in the 35-year history of confrontation against insurgency. Public opinion is extremely upset, devastated and demoralized by what happened.

Colombian Security Analyst, Alfredo Rangel Suarez[1]

What we know is that the offensive was a complete disaster from the military point of view. The army got its butt kicked again. It [Miraflores in August 1998] is the worst in a long string of defeats, and the guerrillas just seem to be getting stronger and stronger while the army just does not seem to be able to turn it around.

United States (US) Official[2]

Beginning in 1996, Colombian security forces faced a series of unanticipated nationwide attacks and suffered increasingly serious defeats.[3] About that same time, the United States had cut most of its security assistance to Colombia because of drug-related and human rights issues. By the summer of 1998—with a new President elected on a peace platform about to assume office and security forces incapable of defeating the left-wing guerrillas—most observers in Colombia and the United States considered the Colombian situation grave. Yet, 10 years later the situation had been reversed and some talked of a possible Colombian Government victory. By 2008, the security forces that had been "not up to the task of confronting and defeating the insurgents" in 1998 dominated the countryside; attacked an enemy reduced in strength by combat actions, desertions, and government programs; and conducted successful hostage rescues and high-value target attacks that demonstrated skillful, professional planning and execution based on actionable intelligence, capable units, and rapid reaction.[4] How this transformation occurred in the midst of fighting is the topic of this study.

What follows is an overview of how the Colombian security forces—at times with and at times without US assistance and advice—grew in size and capability under difficult conditions. The story is shaped, among other things, by conditions in Colombia, by Colombian and US policies, by the actions of the security threats, and by key people. This introductory chapter provides an overview of the situation in Colombia prior to the administration of President Andres Pastrana. The second chapter addresses the interaction between Pastrana's pursuit of peace, the Colombian security forces' adjustment to security and political realities, the actions of multiple threats, the US military reengagement, and the US counterdrug policy from 1998 to 2002. Plan Colombia, with US assistance, became the Colombian approach to address its problems during this period. Chapter 3 covers President Alvaro Uribe's Democratic Security Policy from 2002 to 2006 and its followup 2007 Democratic Security Consolidation Policy. During the Uribe period, Colombian security forces built on the changes begun during the Pastrana era, expanded in size and capability, and became part of a whole-of-government effort to address Colombia's security and governance challenges. The last chapter provides observations on host nation–US interaction based on the Colombian experience that may prove useful to US military personnel performing similar assist-and-support missions in the future.

Colombia: The Setting

Situated at the northwest corner of South America, the Republic of Colombia borders five countries—Panama to its northwest, Venezuela to its east, Brazil to its southeast, Peru to its south, and Ecuador to its southwest. It is the only South American country with coastlines on both the Pacific Ocean and the Caribbean Sea. As the fourth largest country on the continent with 1,141,748 square kilometers, Colombia is roughly the size of Texas, Oklahoma, and New Mexico combined and is divided into five geographic regions.[5] (See Figure 1, Colombian physical geography.) Three parallel north-south mountain ranges—the Cordillera Occidental in the west, the snow-capped Cordillera Central in the middle, and the Cordillera Oriental in the east—make up the Andean region where most Colombians live. The Caribbean coastal region in the north has the major seaports. The underdeveloped lowlands of the Pacific region lie to the west of the Cordillera Occidental. To the east of the Cordillera Oriental are the least developed and least populated areas: the Orinoquia region known for its plains and the Amazon region with its jungles.

The Andean region formed the demographic, economic, and political heart of the country. In 1998, most of the approximately 38 million Colombians—the second highest population in South America—not only lived in the Andean and Caribbean regions, but over 70 percent resided in urban areas. Bogotá, the capital, served as the political center while Medellín acted as an economic center. Infrastructure and government

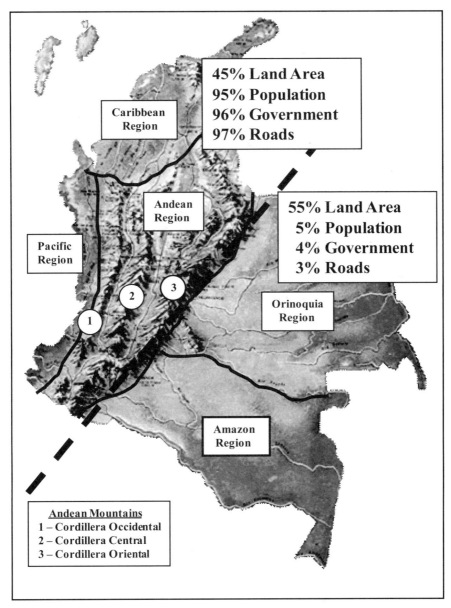

45% Land Area
95% Population
96% Government
97% Roads

Caribbean Region

Andean Region

Pacific Region

55% Land Area
5% Population
4% Government
3% Roads

Orinoquia Region

Amazon Region

Andean Mountains
1 – Cordillera Occidental
2 – Cordillera Central
3 – Cordillera Oriental

Figure 1. Colombian physical geography.

presence focused on the populated areas. The transportation infrastructure consisted of 145,000 kilometers of roads of which no more than 15 percent were paved; over 18,000 kilometers of waterways where riverboats provided the only means of transport in 40 percent of the country; and 3,300 kilometers of unused railroads because of poor security. In addition to traditional agricultural products like coffee, bananas, and cut flowers, the economy depended on industry, services, mining, and oil and gas sectors. Economically, Colombian society divided into an upper class of 5 percent, a middle class of 20 percent, and a lower class of 75 percent.[6] The United Nations (UN) reported that 54 percent of the rural population lived in poverty—almost half of whom lived in extreme poverty—and that 39 percent of the urban inhabitants lived in poverty—15 percent in extreme poverty.[7] Land ownership in the late 1990s reflected this economic disparity—57 percent of land owners held only 3 percent of the productive land while less than 0.5 percent owned 60 percent of the productive land.[8]

Latin America's oldest democracy, described as "procedural" and a "low intensity democracy," focused more on the regular election of officials than on governance of the country.[9] Although the 1991 Constitution, which replaced the 1886 Constitution, and election laws made Colombia more democratic, established judicial and political reforms, strengthened civilian control over security forces, protected individual freedoms, and decentralized governance, Colombia remained a weak state in which "accountability, transparency, corruption, and the ability to extract and distribute resources for national development and protection of human rights and liberties" proved difficult.[10] Acknowledged as "one of the most responsible, accountable, and participatory charters in the region," the 1991 Constitution suffered from being seldom or sparingly applied.[11] Below the national level, 32 departments and a capital district provided regional governance with local affairs handled by elected officials in over 1,000 municipalities.[12] The central government had appointed municipality mayors before 1998 and departmental governors before 1991; afterwards they were elected for 4-year terms. The mayors served a staggered term: 2 years before and 2 years after each 4-year Presidential election.[13] By 1998, over half of Colombia—the large, undeveloped, and unpopulated area east of the Cordillera Oriental where 8 of the 10 departments in the Orinoquia and Amazon regions had not been established until 1991 and where less than 5 percent of the municipalities existed—lacked any permanent government presence and could be reached only by river or air. These ungoverned and uncontrolled areas, along with similar isolated regions in other parts of Colombia, attracted lawless elements and guerrillas.

Multiple Sources of Violence: Drugs, Guerrillas, and *Autodefensas*

Colombia's long history of violence included armed revolts, local violence, banditry, and personal vendettas. Two major civil wars, the

Thousand-Day War of 1899–1902 with over 100,000 dead and *La Violencia* of 1948–58 with over 200,000 killed, failed to resolve the tensions between those political elites who supported a strong central government and those who supported strong regional governments. In the 1960s, Communist and left-wing insurgent groups threatened the state. In the 1980s and early 1990s, drug traffickers became a catalyst for violence and criminal activity. In the absence of Colombian security forces to provide local protection in parts of the country, *autodefensas*—self-defense or paramilitary forces— rose to counter the lawlessness (see figure 2). With over 30,000 homicides a year, of which no more than 10 percent were related to insurgency or armed conflict, Colombia had the highest homicide rate in the world along with the highest kidnapping rate.[14] Demonstrating little concern for life or death, a culture of violence created by historical and current factors—"by the lack of a state monopoly of violence, by the ineffectiveness of the laws . . . by the rules of the narcotics trade . . . and by the class structure of Colombian society"— appeared in many parts of Colombia.[15] By the late 1990s, violence from drug trafficking, insurgency, and *autodefensas*— "Colombia's Three Wars"—threatened the existence of the state from an "unholy trinity of non-state actors . . . perpetrating a level of corruption, criminality, human horror, and internal instability."[16]

Drug Traffickers

For years, Colombia had been the number one producer of coca and of refined cocaine in the world. Heroin and marijuana rounded off the list of homegrown illegal drugs. Crime, violence, and money—lots of money—came with drugs. In the late 1980s and early 1990s, cartels—first in Medellín and then in Cali—dominated the drug business. When the Colombian Government, with US assistance, attempted to destroy the Medellín cartel of Pablo Escobar, the state found itself counterattacked for the first time by drug traffickers. The old "live and let live" or "turn a blind eye" policy died. Assassinations, bombings, threats, and briberies of government civil and security force officials—the "silver or lead" choice between taking bribe money or a bullet—became common. In December 1993, a unit of the Colombian National Police (CNP) killed Escobar and in 1995 the Cali cartel was dismantled.[17] But the demise of the cartels did not end the drug problem; it did not even make it better. In fact, it became more difficult for the CNP to address as smaller, flatter, dispersed, and diverse illegal drug organizations with ties to criminals, insurgents, and paramilitaries replaced the cartels.[18] As long as illicit drugs flourished in Colombia, they would attract, corrupt, and undercut government policies and institutions.

Figure 2. Areas of guerrilla and *autodefensa* activity.[19]

Guerrillas: FARC and ELN

In the early 1960s, numerous insurgent or guerrilla groups inspired by Communist revolutionary movements throughout the world developed in Colombia. While some organizations took their ideological lead from Moscow, others looked to China or Cuba. Although external support remained limited, most groups survived through funds raised from extortions or "taxes" and from kidnappings. In the mid-1960s, the Colombian security forces launched Plan LAZO against Marquetalia and eight other "independent republics"—attacking, reducing, and dispersing the guerrillas into remote areas where they remained a nuisance for almost 20 years. During the 1980s, most of these groups participated in a cease-fire with the Colombian Government in an attempt to negotiate a political settlement. The 1991 Constitution addressed many of their grievances. Several groups demobilized and entered the political process. However, two organizations that failed to reach an agreement—the Revolutionary Armed Forces of Colombia (FARC) and the National Liberation Army (ELN)—grew into significant threats by the mid-1990s. By 1995, guerrillas had established a presence in over 600 municipalities throughout Colombia with their strength concentrated in the newly settled areas of "internal colonization" west of the Cordillera Oriental and of "border colonization" east of those mountains.[20] In 1997, the US State Department designated both the FARC and the ELN as Foreign Terrorist Organizations (FTOs).[21]

In the mountains of southeast Colombia in 1966, the rural-based, Moscow-leaning Colombian Communist Party organized the FARC as its military force from survivors of the "independent republics." A guerrilla since 1949 and a veteran of Marquetalia, Manuel Marulanda—known as *Tirofijo* or "sure shot"—became the FARC's first chief of staff and its eventual commander. In its early years, the FARC's primary effort remained defensive, focusing on survival and gradual expansion—a unit or front in 1969, another in 1971, and a general staff in 1974. Lack of support among the populace made kidnapping the FARC's primary source of funds in 1971. By 1982, the FARC had grown to 17 fronts with about 1,000 armed members.[22]

The FARC leadership held its Seventh Guerrilla Conference from 4 to 14 May 1982. At that meeting, the FARC adopted an 8-year plan to permit a shift from a defensive posture to an offensive strategy that attacked Colombian Military (COLMIL) forces and expanded FARC control throughout the country. The plan included offensive, government, and defense-of-the-revolution phases. To accomplish these tasks, the FARC planned to expand to 28,000 armed members in 48 fronts by 1990. To emphasize this change in direction, it added People's Army (EP) to its name, FARC-EP. The operational concept planned to dominate the eastern slopes of the Cordillera Oriental from its headquarters at La Casa Verde to separate half of the country from the government in Bogotá by employing a "new method of operating"—heavily-armed, mobile guerrilla units in coordinated attacks against isolated military units. To create capable units, the FARC planned to establish training schools and regional training facilities. Previously, fronts had been military units consisting of 50 men in 2 platoons, or guerrillas, divided into squads of 7 or 8. When two or more fronts combined, they became columns. After this conference, fronts assumed regional responsibilities and the task of building additional fronts. The FARC reemphasized its reorientation from conducting minor skirmishes and ambushes to guerrilla warfare at a December 1987 meeting. To finance this effort, specific economic areas were targeted for exploitation: livestock, commercial agriculture, oil, gold, and drugs. The primary sources of funding became extortion and kidnapping. By 1989, the FARC had grown to 7,000 personnel in 27 fronts—a 700 percent increase in 7 years—even though the military portion of its plan had been postponed before implementation because of peace negotiations with President Belisario Betancur's government.[23]

Once it became apparent that the Colombian Government would not agree to its terms, the FARC used the cease-fire to prepare for hostilities. In May 1989, the FARC decided that it would renew its efforts in January 1990. Its 8-year Bolivarian Campaign for a New Colombia, an updated version of its 1982 strategy, envisioned four phases:

(1) 1990–92: Acquire weapons, communications, and aircraft as the FARC grew to 18,000 guerrillas in 60 fronts.

(2) 1992–94: Increase funding for expansion to 32,000 in 80 fronts.

(3) 1994–96: Conduct guerrilla operations with 16,000 and fight the "new method of operating" with 16,000 to overrun military, civilian, and other objectives.

(4) 1996–98: If phase 3 failed, regroup, conduct attacks against infrastructure—roads, electrical systems, communications sites—and plan a second offensive.

To augment its combat forces, the FARC created a local civilian militia. But no plan survives contact with the enemy. After the FARC's refusal to disarm to participate in the Constitutional Assembly election, the COLMIL attacked the FARC headquarters at Casa Verde—Operation *Centauro*—on 9 December 1990, the day of the election. The FARC leadership escaped and 3 weeks later launched OPERATION WASP, the largest series of attacks in Colombia that continued into February 1991. At its Eighth Guerrilla Conference in April 1993, the FARC opted for a war of movement and created a system of seven regional commands or blocs in response to the attack on Casa Verde. Thereafter, the FARC would be decentralized, dispersed, and a more difficult target for the Colombian security forces to attack. Each bloc assumed regional responsibility, consisted of at least five fronts, and raised funds by exploiting local resources. Senior FARC leadership resided in a seven-member Secretariat. The FARC decided to push into southeast Colombia to gain control of the coca-growing area and the borders with neighboring countries. In addition, the FARC conducted information operations to garner support within Colombia, the international community, and nongovernmental organizations (NGOs) operating in Colombia. By 1998, the FARC had grown to well over 10,000 members operating in 61 fronts, 4 mobile columns, 15 mobile companies, and 5 urban fronts. Lacking popular support, what had become the best-organized, equipped, and trained guerrilla force in Latin America relied on kidnapping, extortion, and drugs for funding and on intimidation for recruiting.[24]

The ELN represented a smaller but significant rural-based guerrilla threat in northeastern Colombia. Created in 1964 by university students, liberation-theology-inspired Catholic priests, and oil workers, the ELN looked to Fidel Castro for support and inspiration. It primarily attacked infrastructure—roads, bridges, oil lines, power lines, and dams—and received its financial support from extortion, kidnappings, and aircraft hijackings. Unlike the FARC, the ELN focused on developing a political base before fielding guerrilla units. Like most guerrilla groups, it suffered from internal strife, negotiated with the government at times, clashed with the FARC at times, and aligned with the FARC at other times. Although Colombian security forces had reduced the ELN to 17 members in 1973, by 1998 it had recovered to almost 5,000 members. Right-wing paramilitaries constituted its principal threat.[25]

Autodefensas or Paramilitaries

Armed civilian militias had a long legal and illegal history in Colombia—a result of its record of conflicts between federalism and regionalism and of the absence of security forces in parts of the country. As security threats increased from insurgents and then drug traffickers, the arming of ad hoc, self-defense groups by local leaders became a normal and traditional response. Plan LAZO—developed with US military input—recognized a need for local security forces to remain in areas cleared of guerrillas by military units that moved to other areas. In May 1965, Presidential Decree 3398 authorized the arming of rural civilian self-defense forces under COLMIL supervision. In 1968, this decree became law and served as the legal foundation for organizing and arming paramilitary support for counterguerrilla operations for over 20 years. Local paramilitary groups, working in conjunction with the CNP, had played an important role in the destruction of the Medellín and Cali drug cartels. In response to concerns about ties to drugs and atrocities, a Presidential decree in 1989 made all paramilitary groups, even those that operated under military supervision, illegal.[26] Unfortunately, "making the militias illegal did not uncreate them . . . [but] merely encouraged the creation of multiple groups outside the control of the state" to fill the vacuum created by too few Colombian security forces.[27]

In many rural areas with little or no government presence, paramilitary units with known drug ties and that conducted atrocities garnered more support than did the guerrillas. Why? With no one to turn to for security, one observer noted that the "atrocities of the paramilitaries are not actions of abnormal men, but rather the acts of normal men subjected to and victimized by unremitted violence, who see the disappearance of the guerrillas as the only sure solution to their plight."[28] Because they targeted civilians suspected of supporting the guerrillas, many viewed the paramilitaries as antiguerrilla—not as antistate. They were not officially recognized as a security threat until 1993. Others, particularly human rights organizations and the US State Department, alleged continued COLMIL ties with the paramilitaries from 1993 to 1998 despite Colombian Government denials. Paramilitary activity grew during peace negotiations as government security operations decreased. From 1993 to 1997, the number of these illegal groups increased from about 270 to over 400 with 3,000 to 6,000 members and resulted in increased civilian killings by paramilitaries. (See Table 1, Noncombatant deaths and forced disappearances.) In 1997, Carlos Castano, a paramilitary leader with known long-time drug ties, established the United Self-Defense Groups of Colombia (AUC) to coordinate regional groups in a campaign against guerrilla strongholds. A multiday attack on the population of the municipality of Mapiripan in southern Colombia became just one such notorious event. Despite its illegal status and deserved reputation for atrocity and massacre, paramilitary strength continued to grow as massacres and assassinations targeted guerrilla supporters and displaced populations from parts of Colombia.[29]

Table 1. Noncombatant deaths and forced disappearances[30]

	1993	1995	1996	1997	1998
Guerrillas	28%	38%	36%	23.5%	21.3%
Security Forces	54%	16%	18%	7.5%	2.7%
Paramilitaries	18%	46%	46%	69%	78%

Acknowledging the lack of security in parts of the country, in February 1994 Presidential Decree 356 authorized "special private security and vigilante services" in which anyone with Ministry of National Defense

(MOD) approval could provide for their own security armed with combat weapons. In September, President Ernesto Samper used this decree to create the Community Associations of Rural Vigilance (CONVIVIR), which focused on intelligence reporting for security forces and self-defense by these groups. Within a year, with over 500 units created, almost 10,000 armed men, and acknowledged lack of government oversight, Samper suspended the creation of additional CONVIVIRs. The existing organizations remained legal until abolished by his successor in 1999.[31] The mix-mash of legal and illegal self-defense forces exceeded the capacity of the government to control and failed to provide an effective alternative to the lack of security forces.

Each of the principal nonstate generators of violence—drug traffickers, guerrillas, and *autodefensas*—created formidable difficulties for the Colombian Government and its security forces. In 1998, the government reported that the paramilitaries had committed 70 percent of the human rights violations, the guerrillas 25 percent, and the Colombian security forces 5 percent. In addition to the monies flowing to drug traffickers, funding for the guerrillas and *paramilitaries* that year totaled $551 million from drugs, $311 million from extortion, and $236 million from kidnappings—almost $1.1 billion.[32] Despite these multiple security threats, for years "various Colombian Governments dealt with the problems on a completely ad hoc basis—without a plan; without adequate or timely intelligence; without a consensus among the political, economic, and military elites about how to deal with the armed opposition; and importantly, within an environment of mutual enmity between the civil government and the Armed Forces."[33]

Colombian Security Forces

The 1991 Constitution defined the Colombian security forces—the Public Force—as the "Armed Forces and the national police exclusively." The military organizations—the army (COLAR), navy (COLNAV), and air force (COLAF)—defended "the sovereignty, independence, and integrity of the national territory and of the constitutional order." The CNP—"a permanent armed body of a civilian nature"—maintained "the conditions necessary for the exercise of public rights and freedoms and to ensure that the inhabitants of Colombia may live together in peace." The specifics of administering, raising, training, equipping, and employing each force were established by law. To ensure an apolitical, nondeliberative security force, members could not vote nor take part in any political activities. Constitutional protection for ranks, awards, and pensions prevented their removal except by legal action. Further protection covered crimes committed "while in active service" and "in connection with their service" that would be tried under the Penal Military Code by security force personnel and not by civilian judges under civil law. Finally, the government of Colombia claimed the sole right to have weapons, munitions, and explosives and to give permission to bear arms. Only the permanent armed security forces, operating under government control, had that permission.[34]

Colombian law placed the security forces under the MOD, which had replaced the War Ministry in 1965.[35] (See Figure 3, Ministry of National Defense.) From 1953 to 1991, no civilian served as the Minister of Defense; however, since 1991, all have been civilians. Although civilian-led, the MOD remained a military-staffed organization with few civilian members. The minister had one vice-minister of defense (VMOD). Although adhering to a long tradition of civilian control of the military and the police, security force commanders tended to view civilian control to mean control by the President with whom they often worked directly on security matters rather than control by the Minister of Defense.[36] Within the MOD during peacetime, the military forces focused on territorial defense and the CNP addressed internal security. During internal emergencies, military forces secured infrastructure and conducted operations against the guerrilla bands in the countryside, while the CNP controlled urban areas and municipalities. The capabilities of the security forces to address the internal threats rested on the support or, more accurately, the lack of support of a government and a populace that sought to negotiate, appease, or ignore the sporadic security problem rather than to destroy it. From 1995 to 1998, a US military official saw four Ministers of Defense, four COLMIL commanders, three Colombian Army commanders, six Colombian Army intelligence directors, and noted "in some of the other cabinet ministries it's even worse."[37] The 4-year Presidency, coupled with the frequent turnover of key officials, created a lack of leadership continuity that ensured short-term approaches to long-term problems.

Colombian Military (COLMIL) Forces

In 1951, Colombia created the position of Armed Forces or COLMIL commander, normally filled by a Colombian Army general.[38] More a coordinator than a commander, Colombia had not created a common or joint capability in security matters by 1998. Each military service commander—equal in rank to the COLMIL commander—held both administrative and operational responsibilities for his service.[39] Given the unique history of each service, their different missions, and the scarcity of resources, competition among the armed services was normal and with the national police even more pronounced. COLMIL had roughly 120,000 personnel

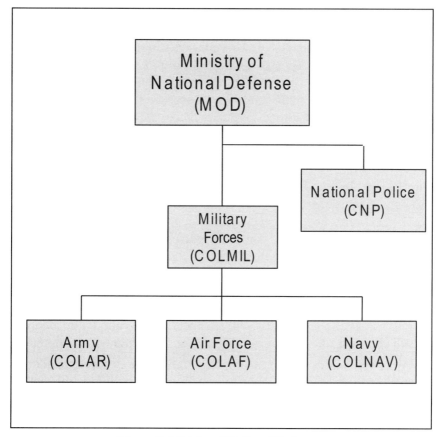

Figure 3. Ministry of National Defense.

in late 1997.[40] That December, the COLMIL commander, General Manuel Jose Bonett, responded to the deteriorating security situation created by what he called collectively the "generators of violence" when he published a military strategy for securing the population and critical resources. However, without a national strategy supported by the government and by a budget, this document proved "nothing more than a statement of principle, a general guideline for action, or an expression of intent."[41] The on-the-ground, limited-resource strategy evolved to protecting the populated areas by establishing a military presence throughout the country by garrisons of small units and by reacting, whenever possible, to guerrilla attacks. This proved "politically sound but militarily disastrous."[42]

In early 1998, the Colombian Army considered itself a professional army generally structured along US lines based on its Korean war service. As in most peacetime armies, its training and education system— considered better "in the garrison and the classroom than in the field" because it lacked "much unit training out in the field"—focused on external, more conventional threats rather than the distraction of counterguerrilla operations.[43] A former US ambassador described the 104,000-man Colombian Army as "basically a barracks military, not . . . organized to go after the guerrillas . . . . [having] some brave and capable people, but . . . strictly a reaction force, and not a very mobile one at that."[44] The soldiers came from two sources: regulars who were conscripts or draftees and professionals who were volunteers. The regulars—80 percent of the Colombian Army—received a basic military training and they formed two types: those with high school diplomas who served 12 months and were exempt from combat duties and those without degrees who served for 18 months. The professionals had completed initial service, volunteered for a 2-year term, and received additional training. They constituted a small, trained combat force. About 20 percent of the army, the 19,900 professionals formed the fighting core of the army while 45 percent—regulars without high school degrees—secured critical infrastructure such as roads, bridges, dams, power systems, and oil pipelines. The remaining 35 percent—those with high school degrees—could not, by law, be sent to combat areas.[45] The officers—all graduates of the military academy—received commissions in infantry, cavalry, artillery, engineer, intelligence, communications, or logistics. Noncommissioned officers—volunteer candidates with a 9th grade education or higher—graduated from a 1½-year noncommissioned officer school. Military occupational specialties for regular soldiers did not exist.[46] As a conscript-based, conventionally-focused force, the Colombian Army dealt with the guerrillas only when the CNP failed to do so by using its small professional units.

Reorganized in 1996, the Colombian Army consisted of five divisions with territorial responsibilities. Each division—commanded by a major general—consisted of two to four territorial brigades and two to three counterguerrilla battalions. The Fifth Division, authorized in 1995, remained a work in progress that included administrative and school responsibilities in Bogotá. Each brigade, commanded by a brigadier general or a senior colonel with a small staff, had a standard organization reminiscent of a US infantry brigade; however, few had all their authorized battalions that lieutenant colonels commanded. (See Figure 4, Colombian Army brigade.) Each of the 16 brigades was authorized 3 infantry battalions—many had 2 infantry battalions assigned and one had 4; a cavalry squadron—7 existed; an artillery battalion—7 existed; an engineer battalion—10 existed; counterguerrilla battalions—1 to 6 assigned to each; a military police battalion—1 or 2 each in 4 brigades; and all had support battalions. The standard infantry battalion consisted of three infantry companies of four platoons filled with regular soldiers and one or more companies of four platoons manned by professional soldiers. Each infantry company—commanded by a captain—was authorized 165 men and the infantry battalion had about 500 men. All battalions had small staffs. Although each battalion in a brigade, regardless of type, had security responsibilities, little institutional motivation existed to engage the guerrillas because any loss of personnel or equipment meant an investigation with probable disciplinary action and any dead guerrilla resulted in a police criminal investigation with possible criminal charges.[47]

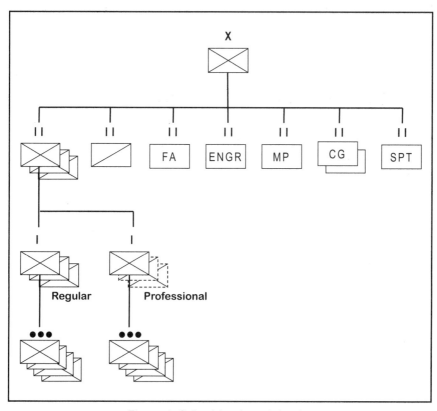

Figure 4. Colombian Army brigade.

To address the guerrilla nuisance—"the traditional enemy of the military,"[48] the Colombian Army created special units of professional soldiers that grew in size and number as the guerrilla threat increased—first with counterguerrilla companies during the 1960s and then with counterguerrilla battalions (BCG) and a mobile infantry brigade (BRIM) in 1990. (See Figure 5, Mobile infantry brigade [BRIM].) In December 1990, 1 BRIM attacked the FARC leadership at Casa Verde. Special taxes

15

raised after the FARC attacks in 1991 permitted the Colombian Army to increase the number of professional soldiers from 2,000 in 1990 to about 15,000 by 1993. In 1991, 2 BRIM was organized and 3 BRIM in 1997. Not truly mobile, a BRIM lacked organic transportation. Mobile merely meant that it had no fixed territorial responsibility; it could move, as required, to threatened areas. Commanded by a colonel, a BRIM consisted of four 360-man counterguerrilla battalions commanded by majors. A battalion had four 84-man companies each with two 41-man platoons. Each counterguerrilla company had half the strength of an infantry company. After 1992, a Counterguerrilla Special Operations Command administered the nondivisional counterguerrilla units. By 1998, the Colombian Army had three BRIMs and a four-battalion Special Forces Brigade created in 1996 with which to respond to countrywide threats. A small army aviation brigade had a limited number of UH-60, UH-1H, and MI-17 helicopters to provide support.[49]

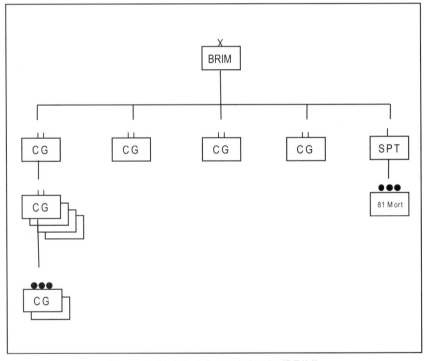

Figure 5. Mobile infantry brigade (BRIM).

As the second largest armed service and less than 15 percent the size of the Colombian Army, the Colombian Navy faced multiple challenges— blue-water operations both in the Pacific Ocean and in the Caribbean

Sea where an island governmental department existed, coastal operations against drug traffickers in both areas, and brown-water or riverine operations along 12,600 kilometers of navigable rivers. Similar to the Colombian Army, the Colombian Navy favored a traditional or professional blue-water approach and resisted involvement in counternarcotics and internal security matters. Its major commands were the Pacific Naval Forces Command and the Caribbean Naval Forces Command. The Colombian Navy consisted of a navy with frigates, submarines, support ships, patrol boats, and fixed and rotary-wing aircraft; a naval infantry organized in 1937—a brigade and its largest component by 1998; a coast guard corps organized in 1979 and equipped with patrol boats; and a riverine force organized in 1956 and equipped with river craft. A naval infantry brigade commanded the naval infantry battalions and riverine units.[50] The Colombian Navy—like most navies of the world—emphasized blue-water, high-technology matters and not counterdrug or internal security issues, which were CNP and Colombian Army concerns.[51]

The smallest armed service with less than 7 percent the manpower of the Colombian Army, the Colombian Air Force possessed a mixture of fixed and rotary wing aircraft to support air supremacy, attack or close air support, air transport, and aerial reconnaissance missions. As with many similar air forces, it had a limited number of aircraft, most not state-of-the-art, from various countries. Air superiority aircraft included French Mirage and Israeli Kfir fighters. Close air support came from Brazilian AT-27 Tucanos and US A-37 Dragonflies, OV-10 Broncos, and AC-47 Spooky aircraft and armed UH-1H Huey and UH-60 Blackhawk helicopters. Spanish CASA aircraft with a mixture of aircraft from other countries provided air transport. Reconnaissance came from helicopter and light aircraft. Organized into numbered regional Air Combat Commands (CACOM), the Colombian Air Force responded to the Colombian Army and CNP internal security requirements when possible if asked. Like many air forces with limited means, the Colombian Air Force was a fair-weather daytime air force.[52]

Colombian National Police (CNP)

As the second largest security force in Colombia in 1998, the CNP numbered about 90,000 members commanded by Major General Rosso Jose Serrano. A long history of mistrust, dislike, and misunderstandings with the COLMIL and the bureaucratic battles over limited resources made cooperation difficult. Responsible for law enforcement throughout the country, the CNP had diverse responsibilities requiring urban, rural, antinarcotics, criminal investigation, antikidnapping, customs, highway patrol, intelligence, and special operations units. With US counterdrug assistance provided in the mid-1990s, the CNP became heavily armed for a national police force and its antinarcotics police aviation assets—56 helicopters and 17 fixed-winged aircraft along with 15 US-provided spray aircraft and 12 helicopters—provided a mobility envied by the COLMIL. Although counterdrug efforts made the CNP targets for guerrillas and criminals, US counterdrug assistance intensified the tensions between the COLMIL and the CNP. A senior COLMIL officer complained to an American, "We now have two armies and you are responsible—you always support them—you treat them better than us!" Yet, in many isolated municipalities, a 40-man police unit provided law and order. Increasingly, guerrilla attacks targeted these isolated small units.[53]

Involvement in the internal security of the nation proved difficult for the COLMIL, as it does for most militaries. First, not being at war or under a constitutionally sanctioned "state of internal commotion" legally declared by the President, normal laws applied. In peacetime, only the CNP could arrest or detain a suspect. When anyone died during a small-unit military operation, operations ceased, the "crime scene" secured, and a police investigation conducted to confirm or to deny criminal activity. Seldom did a judicial technical police investigator accompany a military unit on counterguerrilla operations. Often it took hours, and sometimes days—depending on the remoteness of the location—for the investigator to arrive. Under these conditions, only limited tactical successes proved possible. A senior military officer noted that the laws seemed more suited to Denmark than to the conditions in Colombia. Second, the normal actions of combat operations—detentions, killings, and other actions in a confusing, stressful environment—meant possible liability, potential criminal charges, and accusations of human rights violations. For these reasons, COLMIL officers strongly defended their Constitutional Article 221 protection of being judged by the military for accused crimes committed during an act of service. What many in the security forces considered an essential legal protection, many in the human rights community viewed as impunity from prosecution.[54]

Colombian–United States Relations

From just prior to World War II, Colombia and the United States maintained close diplomatic and military relations. In 1939 the United States established a military mission in Bogotá.[55] The only Latin American country to send military forces to the Korean war, Colombia provided an infantry battalion and a frigate shortly after the outbreak of hostilities. These United States-trained-and-equipped Colombian units served with

US divisions and naval forces.[56] In 1956, US ranger-qualified personnel assisted in the establishment of a *Lancero* school, the first such course in South America.[57] From a Special Survey Team in 1959 to the Yarborough Team visit in 1962 through the execution of Plan LAZO in the mid-1960s, close cooperation and consultation between Colombia and the United States continued.[58] Colombia participated in the United Nations Emergency Force I in Egypt during the 1956 Suez crisis and in the United States-sponsored Multinational Force and Observers in the Sinai where it provided a battalion along the Israeli–Egyptian border since 1981.[59] After President George H.W. Bush declared "war on drugs" in 1989, Colombian and United States activities became increasingly intertwined.

Country Team and the Counterdrug Policy

In 1994, Myles R.R. Frechette arrived in Colombia as the ambassador for the largest US Embassy overseas. Principal members of his country team included the Department of State's (DOS) Narcotics Affairs Section (NAS), the Department of Defense's (DOD) Military Group–Colombia (MILGP), US Agency for International Development (USAID), and law enforcement agencies that included the Drug Enforcement Agency (DEA), Department of Justice (DOJ), Federal Bureau of Investigation (FBI), US Coast Guard (USCG), and US Customs Service. US policy concentrated on what was called the four Ds: democracy, development, drugs, and *derechos humanos* (human rights).[60] After the election of President Samper in 1994, allegations of a $6 million payment from the Cali drug cartel brought an immediate and harsh US reaction in what would become a 4-year adversarial relationship over corruption and the democratic process, ineffective alternate crop development and drug eradication efforts, resistance to the US-financed counterdrug program, and an atrocious human rights record. For Americans, illegal drugs—the principal reason for engagement in Colombia—generated the violence and corruption that made Colombia a potential quagmire.

Two ghosts from the recent past influenced United States policy in Colombia—Vietnam and El Salvador. From Vietnam came the imperative not to become involved in a counterinsurgency and from El Salvador came the necessity of emphasizing human rights. This meant maintaining a strict policy distinction between counterdrug and counterinsurgency efforts— no matter how irrelevant to the actual situation or how beneficial to the institutions involved—as well as a spotless human rights record. In early 1992, Colombia and the United States agreed that the CNP would lead the counterdrug effort—the NAS working with the antinarcotics police directorate (DIRAN). Counterdrug funding for the COLMIL ceased.

19

This suited both countries. For the Americans, the CNP had proven a more cooperative partner and its DIRAN focused on counternarcotics. The COLMIL, on the other hand, had a poor human rights record and had been uncooperative—unable, if not unwilling, to maintain a strict distinction between counterdrug and counterinsurgency operations that the Americans required for counterdrug assistance. For the Colombians, the DIRAN became the focal point for the US-supported counterdrug effort freeing the Colombian security forces—COLMIL and CNP—to address the spreading violence generated by criminals, guerrillas, and *autodefensas*. This decision reinforced the COLMIL perspective that drug trafficking was a crime, thus a police task, while guerrillas were a military threat, thus a military task.[61] The COLMIL resisted what it considered an artificial distinction between counterguerrilla and counterdrug operations as it confronted what a COLMIL commander called the "narco-guerrilla" threat.[62] Many Americans viewed the Colombian emphasis on fighting guerrillas rather than counterdrug operations as "not the best way to resolve its security problems, not the least because of opposition in Washington to counterinsurgency."[63] These different views of the security problem created miscommunication, misunderstanding, confusion, and frustration for Colombians and Americans.

The US counterdrug strategy in Colombia had three goals: (1) to take down the drug trafficking leadership, (2) to reduce the amount of drugs through eradication, and (3) to strengthen legal institutions. The first became the focus of the CNP and DEA; the second the CNP and NAS; and the third the government of Colombia and several US agencies—DOS, DOJ, FBI, and USAID. The key programs focused on eradication of coca plants and destruction of drug laboratories, aerial and maritime interdiction, judicial reform, prevention of money laundering, drug awareness education, and, because nothing could be done instantly, development of infrastructure to support these programs.[64] An Air Bridge Denial Program, consisting of several radar sites in Colombia, identified suspect aircraft that the Colombian Air Force attempted to intercept in its airspace. The Colombian Navy—in coordination with the USCG, DEA, DOD, and US Customs Service—participated in interdiction efforts in its territorial waters. Aerial spraying, initially resisted by the Colombians, began in parts of Colombia in the early 1990s. The counterdrug program suffered from high expectations and poor outcomes. The American focus on results created short-term programs that took time to develop, effort to coordinate, capacity to execute, and experience to make effective. Some believed the "US has been and is still today trying to fight the drug war 'on the cheap.'"[65] Others claimed the counterdrug program had become the guerrillas' "best friend" because eradication success in the neighboring countries of Peru and Ecuador had pushed coca production into southern Colombia and because their major competition—the drug cartels—had been dismantled.[66] From 1990 to 1998, the over $500 million spent on antinarcotics eradication brought no reduction in coca production—in fact, it rose by 50 percent.[67]

In 1998, Colombia reportedly had the "worst human rights record in the hemisphere." The indexes—political killings, disappearances, massacres, forced displacements—reflected a country struggling with multiple sources of violence.[68] Although the *autodefensas* and the guerrillas committed the majority of the violations and the security forces' violations had declined over time, the COLMIL—particularly the Colombian Army—struggled to counter the negative image created by FARC information operations, the critical reports of NGOs, and DOS human rights reports. Responsible for security throughout the country, the security forces received criticism not only for committing violations, but also for failing to prevent abuses. In 1997, when the guerrillas killed over 200 candidates for elected office and intimidated over 2,000 to withdraw and when the paramilitaries committed the majority of the massacres and political killings—to include their brutal attack on Mapiripan—the security forces and the judicial system continued to receive harsh criticism.[69] Almost 75 percent of all crimes were not reported and less than 3 percent of all crimes were punished—the indicators of a broken legal system. A DOS report addressed concerns about military impunity, about operating under emergency or "state of internal commotion" decrees in parts of the country for 36 of the past 44 years, and about the "persistent, unofficial, emphasis . . . on body count as a means of assessing field performance . . . [that makes it] a main contributing cause of violations."[70] Past ties to self-defense groups, the inability or unwillingness to attack the growing paramilitaries, the repeated reports of security force personnel collusion, the inverse relationship of paramilitary violence increasing as security force abuse dropped (see figure 2), and the continued, although reduced, violations, made it impossible for the Colombian Army to overcome its poor human right reputation.

Colombian–US relations deteriorated during the Samper presidency. Corruption, counternarcotics, human rights, different priorities, and personalities all played a part as both countries pushed the limits of what the other proved willing or able to do. In an attempt to energize Colombian counterdrug efforts in 1995, President William Clinton granted Colombia a vital national interest certification waiver that placed restrictions on counterdrug funding and requirements for future certification—a strong indication of US dissatisfaction. Colombian failure to respond to these concerns resulted in decertification in 1996—a first for a counternarcotics partner and aid recipient—and again in 1997. Colombia received a vital national interest waiver in 1998. The specific factors cited for decertification included governmental corruption undermining law enforcement and the judiciary, poor prison security, failure to safeguard US investigative information, failure to implement an extradition treaty signed in 1979, and inability to reach agreement on herbicides for eradication. Decertification brought Congressionally mandated termination of most foreign assistance programs.[71] Ironically, the CNP counterdrug funding and humanitarian aid continued while the COLMIL and other funding ended. In 1996, Congress passed the Leahy amendment that mandated a human rights certification by the Secretary of State for any foreign military unit receiving US counterdrug assistance. If any "credible evidence that . . . [a] unit has committed gross violations of human rights" existed, then funding was prohibited. Since "unit" was not defined by law, an elaborate, rigorous, time-consuming vetting process for all members of a military unit arose that frustrated many in the COLMIL, particularly because it did not apply equally to the CNP.[72]

Limited US assistance to Colombia during this decertification period reflected the US counterdrug policy, not—from the Colombian perspective—the realities. (See Table 2, US assistance to Colombia, 1995–1998.) Although the MOD received conditional counterdrug funding for CNP antinarcotics units, the COLMIL received none. Not only had US aid been reduced, the utility of much of what was provided for counterdrug programs from the DOD drawdown program suffered from poor planning, lack of coordination, and misunderstandings within US Government agencies and with Colombia. Excess DOD nonlethal assistance for counterdrug programs included protective and utility personal equipment, such as uniforms and helmets; night vision systems; communications, navigation, radar, and photo equipment; spare parts—components, attachments, accessories, hardware, and software—for aircraft and patrol boats; and river boats.[73] A special 1997 drawdown program included five 10-passenger C-26 aircraft added by the National Security Council—two for CNP and three for COLMIL, 12 UH-1H helicopters for CNP, and a landing craft and 6 river patrol crafts for the Colombian Navy. Each of these packages raised Colombian and United States expectations for increased counterdrug capacity. However, Embassy and Colombian officials noted the aircraft had not been configured for CNP and COLMIL surveillance needs, which would require modifications costing at least $3 million each. The DOS agreed to provide $1 million to upgrade the two CNP aircraft, but nothing for the COLMIL aircraft. The 12 helicopters delivered to the CNP in May 1997 had, on average, less than 10 hours of flight time remaining before requiring major maintenance. In July with only two helicopters operational, the CNP announced its intention to use these helicopters for spare parts to maintain its 38-helicopter fleet. In August, the DOS agreed to assist in making the helicopters operational. When a promised excess $1.5 million landing craft could not be found, the COLMIL rejected a smaller vessel as failing to meet its needs. To become operational, the six river patrol craft—decommissioned in 1993—required $600,000 in maintenance just to be of marginal utility and when delivered, many lacked radios and other critical equipment. Embassy and DOS personnel had learned the pitfalls of using DOD drawdown assistance—excess, obsolescent, or obsolete equipment—for the counterdrug program.[74] The General Accounting Office (GAO) reported not only that this package failed to include "sufficient information on specific Colombian requirements, the ability of the host country to operate and maintain the equipment, or the funding necessary . . . to support it"; that items "did not meet the priority needs of the Colombian police and military"; and that "much of the equipment was not operationally ready for use," but that it had been delayed over 10 months because of delays by DOS in providing the Embassy guidance on human rights vetting procedures and by negotiations with the Colombian Government.[75] US assistance did not always assist. In fact, hastily conceived, uncoordinated, or poorly executed programs made the situation worse, strained tenuous relationships, and encouraged distrust.

Table 2. US assistance to Colombia, 1995–1998[76]

Dollars in millions	1995	1996	1997	1998	TOTAL
Economic	**1.3**	**.6**		**.5**	**3.4**
USAID				.5	.5
Food Aid Grants					
Other	1.3	.6			2.9
Counternarcotics	**18.5**	**22.6**	**56.5**	**99.1**	**196.7**
DOS International Narcotics Control	16	16	33.5	46.3	101.8
DOS Air Wing	2.5	6.6	10.9	37.8	57.8
DOD Sec 1004—CD/Police			10.3	11.8	22.1
DOD Sec 1033—Nonlethal Riverine				2.2	2.2
Administration of Justice			1.8	2	3.8
Military	**10.6**			**.2**	**10.8**
IMET	.6			.2	.8
FMF Grants	10				10
Drawdowns		**14.5**	**9.4**	**18.8**	**42.7**
DOD Sec 506—Nonlethal Excess		14.5	9.4	18.8	42.7
TOTAL	**30.4**	**37.7**	**66.6**	**119.6**	**256.3**

USSOUTHCOM

By 1998, US Southern Command (USSOUTHCOM)—the lowest priority and smallest American military combatant command—had moved its 1,100-person headquarters from Panama into temporary facilities in Miami, Florida. (See Figure 6, USSOUTHCOM.) General Charles E. Wilhelm, commander since 1997, oversaw an area of responsibility that included 32 countries—19 in South and Central America and 13 in the Caribbean and the contiguous waters. With 7,500 military personnel assigned, USSOUTHCOM had five component commands—in Panama: 1,500 soldiers in US Army South (USARSO) at Fort Clayton and US Air Force South (AFSOUTH) at Howard Air Force Base; in Puerto Rico: US Navy South (USNAVSO), US Marine Force South (USMARFORSOUTH), and US Special Operations Command South (USSOCSOUTH) at Roosevelt Roads; and two joint task forces: a 1,000-person Joint Task Force *Bravo* (JTF *Bravo*) at Soto Cano Air Base in Honduras and a Joint Interagency Task Force–South (JIATF-S) formed from a dozen government agencies and departments at Key West, Florida.[77] USSOUTHCOM pursued a "strategy of cooperative regional peacetime engagement." Wilhelm understood that each country in his area of responsibility was unique in its "level of prosperity, stability and history" and that the region ran more on "handshakes and personal relationships" rather than on formal agreements. As its responsibilities had increased, USSOUTHCOM resources had decreased—a personnel reduction of 50 percent from 1994 to 1997 and a foreign military financing (FMF) reduction from $221 million to $2 million from 1991 to 1997. As a result of these reductions, National Guard and Reserve units grew in importance—completing over 40 percent of USSOUTHCOM deployments in 1997. USSOUTHCOM focused on security assistance—particularly FMF and International Military Education and Training (IMET) Programs, humanitarian assistance, civic assistance, human rights, and counterdrug programs.[78] Relocated from Panama to Fort Benning, Georgia, in 1984, the US Army School of the Americas provided most Spanish-language IMET courses.[79]

In Colombia, MILGP–Colombia—commanded by an Army colonel with regional experience, special operations or foreign area background, and Spanish-language skills—constituted a security assistance organization staffed by joint personnel, many serving on temporary duty. As a component of the country team, the MILGP worked for the ambassador.[80] DOD support to counterdrug operations created an "unprecedented mixing of law enforcement and military missions" that had unintended consequences.[81] Lacking a national police and a military involved in internal security, the United States not only had no police organization to align

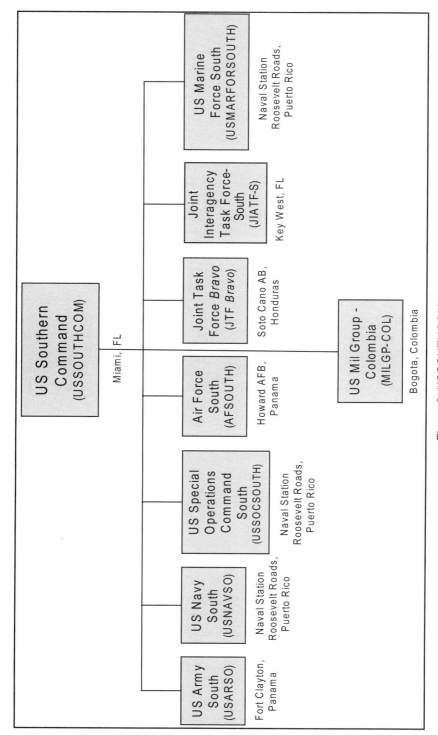

Figure 6. USSOUTHCOM.

25

with the CNP, but it had limited experience with, and little understanding of, the national police-internal security military structure common to most countries in the USSOUTHCOM area.[82] US support of the CNP counterdrug program created a distortion or "institutional imbalance and exacerbated rivalry" between the CNP and the COLMIL. In addition to militarizing the CNP, it reinforced COLMIL attention on counterguerrilla and infrastructure protection—rather than counterdrugs—as the "core of their institutional mission."[83]

Just as the Samper administration resisted American pressures, "Colombian strategies, military institutions and practices with roots deep in a national political, social, and military culture" resisted "alteration despite MILGP-suggested changes."[84] Constrained by decertification restrictions and human rights concerns, the MILGP had little to offer the COLMIL, even if it had been inclined to listen, other than pressure to improve its human rights record. The Joint Combined Exchange Training (JCET) Program constituted about the only training opportunity available for the COLMIL. Created in 1991 to provide training for US Special Operations personnel to gain "proficiently in their mission-essential tasks . . . language training, cultural immersion, and knowledge of the local terrain and weather," JCET permitted training with host nation units without the funding or other constraints of assistance programs because this program trained US personnel by working with a host nation unit.[85] JCET training involved a small group of Americans training with a Colombian company or personnel from a battalion for a limited period. The number of Colombians trained in basic skills by JCET missions remained insignificant. When the press alleged that JCET missions violated US policy in Colombia, the Embassy responded, "the few JCETs . . . were consistent with foreign policy objectives" and "because only one or two took place each year, they did not have a major impact of the achievement on US goals."[86]

Road to the Brink of Disaster

In 1996, the FARC transitioned from isolated guerrilla attacks to phase 3, mobile warfare, which concentrated trained, well-armed units for massed attacks against not just isolated CNP stations, but COLMIL units. In April 1996, a force composed of three fronts ambushed a 49-man 6-vehicle military convoy patrolling the Trans-Andean oil pipeline in southern Colombia near the border with Ecuador. Outnumbered, the convoy lost 31 killed and 18 wounded. The Colombian Army credited the attack to "bandits from the cartel of the FARC."[87] At the end of August, the FARC launched an offensive with 22 simultaneous attacks—many multifront—against isolated police and military targets in 12 departments. A Colombian Army officer noted that the assaults "were carried out in different directions, each with great initial intensity that caused disorientation among the unit commanders, who, seeing their troops in imminent danger, caused even greater confusion in army headquarters as it tried to respond across the length and breadth of the country."[88] At Las Delicias, a small, remote river post and counterdrug base in the department of Putumayo near the Ecuador border, an 800-man FARC unit brutally attacked and overran a regular infantry company base for the first time. Overwhelmed by numbers and over 350 mortar rounds and gas-cylinder bombs, the post was destroyed and the garrison killed, wounded, or captured.[89]

The FARC attacks employing massed, well-armed units presented a major threat to small, isolated CNP and COLMIL garrisons. The CNP director noted that the "civil war which we had hoped was a thing of the past is rapidly reviving."[90] Many Colombians saw the August attacks, which coincided with peasant protests against coca eradication, as a reprisal for the US-sponsored counterdrug program. One analyst offered, "We are seeing a military escalation in the war against drugs. We are mixing the guerrilla conflict . . . with the antidrug conflict with the worst possible consequences."[91] In the aftermath of Las Delicias, President Samper considered setting up a FARC-demanded demilitarized area, but the discussions stalled. By the end of the year, CNP counterdrug aircraft had been attacked at an unprecedented rate and FARC attacks and peasant anticoca eradication demonstrations had occurred in 15 departments. When kidnappings reached a rate of one every 6.5 minutes, the security forces created special police and military antikidnapping and antiextortion units—Groups of Action Unified for the Liberation of Persons (GAULA). Facing a deteriorating security situation, Colombian frustration with the United States separation of counterdrug and counterinsurgency assistance led a former senior official to declare what many Colombians believed: "We can't accept such a stupid dichotomy that the two have nothing to do with each other. It is absurd."[92] While American officials believed that the COLMIL did not have the capability to defeat the FARC, they thought that the FARC lacked both the political support and the military power to threaten the Colombian Government. Besides, US policy and legislation forbade support for Colombian counterinsurgency operations.

Conditions in Colombia continued to deteriorate in 1997 as the government struggled to address the guerrilla threat while at the same time facing US counterdrug pressure. In February, the US decertified Colombia for a second year—American counterdrug programs taking precedence over any security concerns. Having withdrawn its security forces from remote parts of the country, the Colombian Government saw in April the establishment of the AUC—an umbrella organization for the growing paramilitary forces. Fundamental disagreements between President Samper, who pursued negotiations with the FARC, and his COLMIL commander, who opposed negotiations, refused to provide the FARC a demilitarized area, and defended the CONVIVIR paramilitaries, led Samper to replace his COLMIL commander with General Bonett in June.[93] In the fall, the FARC sought to disrupt the municipality and department elections through a campaign of assassination and intimidation. Over 200 candidates died and almost 2,000 withdrew from the election. Despite these efforts, the elections occurred without major incident showing a public rejection of the FARC. At the end of the year, the FARC launched a widespread "Black December" offensive in which a force of up to 200 guerrillas from two fronts overran a 32-man Colombian Army platoon occupying a remote mountaintop communications relay site at Cerro de Patascoy near Ecuador. The outnumbered teenage conscripts quickly fell to the combat experienced and more heavily armed guerrillas.[94]

The Americans continued to work their counternarcotics efforts. In late 1996 and early 1997, the United States pushed for increased counterdrug operations in southern Colombia by offering the nonlethal, excess DOD package discussed earlier. The CNP, agreeing early, received its helicopters in May 1997. The COLMIL, however, failed to reach an agreement with the Embassy on human rights vetting requirements until August 1997, after which both the Colombian Air Force and Colombian Navy quickly complied. The Colombian Army—bitter about decertification, human rights vetting, and counterguerrilla restrictions—rejected the assistance.[95] Adverse publicity for this nonlethal package raised a so-called "difficult question of arming a military well known for abusing human rights, especially when the material might be used to put down insurgency among the military's enemies rather than to fight drugs." Although during a visit former USSOUTHCOM commander and "drug czar" Barry McCaffrey often used the COLMIL expression "narco-guerrillas"—a term that Ambassador Frechette repeatedly attacked as inaccurate and a COLMIL funding ploy, McCaffrey stated clearly: "Let there be no doubt: We are not taking part in counterguerrilla operations."[96] In October, the DOS issued its first FTO list that included both the FARC and the ELN among the 30 groups so designated, but it acknowledged no connection between terrorism and illegal drugs.[97]

In November, the US Defense Intelligence Agency (DIA) concluded that within 5 years the COLMIL faced defeat by drug-financed Marxist guerrillas unless the government regained political legitimacy and the

COLMIL was "drastically restructured." The DIA estimated guerrilla strength at more than 20,000—15,000 in FARC and 5,000 in ELN. About two-thirds of the FARC units and one-half of the ELN units had ties to illegal drugs that provided tens of millions of dollars each year. These funds permitted the acquisition of armaments, which included heavy weapons, a few surface-to-air missiles, and small airplanes to move personnel and munitions within Colombia. Most Colombian and American officials acknowledged that the guerrillas controlled over 40 percent of Colombia. In 10 years, the number of municipalities in which the guerrillas maintained a presence had risen from 173 to almost 700. To combat the guerrillas, the COLMIL—described as "inept, ill-trained, and poorly equipped"—had roughly 120,000 personnel, of which only 20,000 professionals were equipped for combat duty.[98] Although this report emphasized the guerrilla threat and its connections to narcotics, most US officials—particularly those in Congress—continued to view the problem in Colombia as counterdrug and not counterguerrilla rather than counterdrug and counterguerrilla. Colombian security officials leaned toward a counterdrug and counterguerrilla interpretation to gain access to US funding, but they saw counterdrug as a CNP task and counterguerrilla as a Colombian Army function.

From 1996 to early 1998, FARC tactical successes came from carefully planned simultaneous massed attacks throughout the country by specially trained units against weak and isolated CNP or COLMIL outposts. This changed the first days of March near El Billar on the Caguan River in the southern department of Caqueta when 52 BCG—a battalion in the newly raised 3 BRIM—moved to disperse a large FARC concentration in a remote area. On 1 March, 153 men from parts of three counterguerrilla companies moved into a U-shaped ambush established by over 450 guerrillas from two FARC fronts and an elite special column. The FARC units surrounded this force and attacked it for 3 days—killing 62, capturing 43, and dispersing the remainder. Overcast skies limited air support and reinforcement. For the first time, an elite counterguerrilla unit had been defeated by FARC guerrillas in open warfare.[99] Colombian defense analyst Alfredo Rangel declared, "This is without a doubt the biggest defeat in the 35-year history of confrontation against insurgency."[100] The COLMIL provided an 80-page response to legislative inquires after this disaster blaming the defeat on intelligence failure and noting the limitations of the Colombian Army: the professional 20 percent fighting, another 45 percent providing infrastructure security, and the remaining 35 percent excluded from fighting by law. It proposed an increase to 60,000 professionals, which would cost $1.6 billion compared to the current budget of

$1.05 billion. The report stated, "In view of the territories involved, the population and subversive force that we confront, it is clear our forces are insufficient."[101] Whether the government would provide additional resources or would seek a political settlement depended on the results of the Presidential election in progress.

That same month, General Wilhelm told a Congressional commit-tee he was encouraged Colombia had received a national interest waiver and acknowledged that a new US Ambassador, Curtis Kamman, had been appointed. Wilhelm described Colombia as "ill-prepared to effectively counter these threats, due in part to weak national leadership and an over-loaded, often corrupt, judicial system, and in part, due to the ineffective-ness of its security forces." He expressed "little cause for optimism" that the COLMIL could "reverse the erosion of government control over the outlying departments." USSOUTHCOM analysis of Colombia focused on countering what he called an "alliance of convenience between the narcotraffickers and insurgents." Wilhelm considered the primary weak-ness of the COLMIL as their "inability to see threats, followed closely by their lack of competence in assessing and engaging them." To improve the COLMIL capacity and professionalism, the MILGRP identified seven areas for improvement: mobility, direct attack capabilities, intelligence systems, night operations, communications systems, sustainment, and operating on rivers and in coastal regions. Wilhelm then noted the lack of a "national strategy that states that it is an objective of the government of Colombia to defeat the insurgency or narcotrafficking."[102] Testifying at the same hear-ing, CNP Director Major General Serrano noted the nexus of the guerrillas and the narcotraffickers—saying "this is true war"—and promised to "fight shoulder-to-shoulder with DEA, with CIA [Central Intelligence Agency], with FBI and, of course the State Department." He made no reference to COLMIL.[103] Wilhelm followed with a letter to General Bonett in which he said, "at this time the Colombian Armed Forces are not up to the task of confronting and defeating the insurgents" and that as the most threatened country in his area of responsibility, the COLMIL was "in urgent need of our support."[104] Despite the dire military need, Ambassador Kamman emphasized, "We are not in a position to hand out important military aid [to the military]. We want to strengthen . . . [COLMIL] in any areas where we have got the resources and the legal basis to do so." Decertification had limited resources and US law prohibited military aid to "those units reputed to have committed human rights abuses."[105]

After El Billar, the Colombians began to work more closely with the Americans. In response to the El Billar disaster and to continued US pressure for counterdrug operations in southern Colombia, the COLMIL established Joint Task Force–South (JTF-S) organized around 2 BRIM at Tres Esquinas Air Base in the department of Caqueta to keep the FARC from establishing a base of operations to control the drug trade and from establishing an "independent republic."[106] Under strong US pressure, in May COLMIL disbanded its 20th Intelligence Brigade—notorious for human rights abuses and ties to the *autodefensas*—and separated its intelligence gathering function from its operations function. A new Army Military Intelligence Center (CIME) centralized army intelligence gathering and worked through Army Military Intelligence Regions (RIME) collocated at each division. Time would determine the effectiveness and consequences of these actions.[107]

Following the hostile relationship of the Samper presidency, many hoped for a closer relationship between Colombia and the United States after the 1998 election. Conservative Andres Pastrana, son of a former President, ran on a platform calling for peace negotiations with the guerrillas. Winning a runoff election that had an unprecedented 60 percent turnout by the closest margin ever in Colombia, Pastrana began talks with the FARC before becoming President. Among the guerrilla demands was a FARC-controlled area free from Colombian security force presence from which to negotiate.[108] Despite high hopes for peace, expectations remained low. No one knew if after 7 August Pastrana could negotiate a peace settlement, if the guerrillas would participate in good faith, or if the security forces would maintain their own. The guerrillas provided a hint in the first week of August, when the FARC and ELN launched a coordinated offensive by making 42 attacks in 14 departments. They used mortars and devastating gas-cylinder bombs in the larger attacks. The largest attack took place against a key antinarcotics police base for refueling aircraft for coca eradication spraying operations and against a Colombian Army company garrisoned at Miraflores in Guaviare department. Five FARC units formed a 1,200-guerrilla assault force. Expecting an attack, the 120-man infantry company dispersed to platoon positions that the guerrillas quickly overwhelmed. The 80-man police unit resisted until it ran out of ammunition and then surrendered. Total security force losses at Miraflores included 30 dead, 50 wounded, and 100 captured.[109] During this offensive, Colombian security forces fought well—often under adverse conditions. But, the guerrillas had the initiative and the demonstrated ability to destroy isolated security force outposts. The guerrillas sent a clear message to Pastrana of their military strength and of the Colombian security force vulnerability.

Notes

1. Laura Brooks, "Colombian Military Is Called To Account: Rebels Outwitted Forces, Critics Say," *Washington Post*, 8 March 1998, http://proquest. umi.com/pqdweb?index=5&did=27000487&SrchMode=2&sid=3&Fmt=3 &VInst=PROD&VType=PQD&RQT=309&VName=PQD&TS=1241191 019&clientId=5904 (accessed 5 November 2008).

2. Laura Brooks, "Debacle in Colombia Shows Rebel Strength," *Washington Post,* 6 August 1998, http://proquest.umi.com/pqdweb?index=18 &did=32661376&SrchMode=2&sid=17&Fmt=3&VInst=PROD&VType=PQ D&RQT=309&VName=PQD&TS=1236966447&clientId=5904 (accessed 5 November 2008).

3. For this study, the term "security forces" refers to the Armed Forces and the national police. The term "military forces" means the three Armed Forces— Army, Navy, and Air Force.

4. USSOUTHCOM commander quoted in 6 April 1998 letter to COLMIL commander, Douglas Farah, "Colombian Rebels Seen Winning War; US Study Finds Army Inept, Ill-Equipped," *Washington Post,* 10 April 1998, http:// proquest.umi.com/pqdweb?index=9&did=28551847&SrchMode=2&sid=2&F mt=3&VInst=PROD&VType=PQD&RQT=309&VName=PQD&TS=1237824 894&clientId=5904 (accessed 5 November 2008).

5. Land area data taken from briefing slide, Republic of Colombia, "Colombia: Building A Path Toward A New Horizon," 2008. Other comparisons can be made to the size of France, Spain, and Portugal or of Texas, Oklahoma, Arkansas, and Louisiana or of all the US states east of the Appalachian Mountains from Maine to Florida (minus West Virginia). The point is Colombia is not small.

6. For general information, see various sections in *Jane's Sentinel Country Risk Assessment,* http://sentinel.janes.com/docs/sentinel/SAMS_country.jsp? Prod_Name=SAMS&Sent_Country=Colombia& (accessed 20 November 2008), or Central Intelligence Agency, *The World Factbook,* https://www.cia.gov/library/ publications/the-world-factbook/geos/co.html (accessed 7 August 2008).

7. Nina M. Serafino, "Colombia: Conditions and US Policy Options," Congressional Research Service Report, updated 12 February 2001 (Washington, DC: Library of Congress), 5.

8. William Aviles, "Paramilitarism and Colombia's Low-Intensity Democracy," *Journal of Latin American Studies* 38 (2006): 391.

9. Aviles, "Paramilitarism and Colombia's Low-Intensity Democracy," 380.

10. Max G. Manwaring, *Non-State Actors in Colombia: Threats to the State and to the Hemisphere* (Carlisle Barracks, PA: Strategic Studies Institute, May 2002), 76–77.

11. James L. Zackrison, "Executive Summary: Workshop on Security Issues in Colombia," in *Crisis? What Crisis? Security Issues in Colombia*, ed. and trans. James L. Zackrison (Washington, DC: Institute for National Strategic Studies,

National Defense University, 1999), http://www.ndu.edu/inss/books/books%20-%201999/Crisis%20What%20Crisis%20Eng%20Oct%2099/cris3.html (accessed 11 February 2009).

12. The number of municipalities grew between 1998 and 2008 to almost 1,100. Municipalities may have several towns. They are similar to counties in the United States.

13. Nazih Richani, "Caudillos and the Crisis of the Colombian State: Fragmented Sovereignty, the War System and the Privatisation of Counterinsurgency in Colombia," *Third World Quarterly* 28, no. 2 (2007): 407. Information on department governors came from a presentation by a Colombian official.

14. Colombia's homicides were twice the number of the United States, which had a population seven times larger. For the United States to have had the same rate as Colombia would have meant well over 210,000 homicides. Angel Rabasa and Peter Chalk, *Colombian Labyrinth: The Synergy of Drugs and Insurgency and Its Implications for Regional Stability* (Santa Monica, CA: Rand, 2001) 17; Dan D. Darrach, "Workshop Session," in Zackrison, ed., *Crisis? What Crisis?*

15. Peter Waldmann, "Is There a Culture of Violence in Colombia?" *Terrorism and Political Violence* 19 (2007): 604–606.

16. Manwaring, *Non-State Actors in Colombia*, 69.

17. For an insight into the challenges posed by Escobar, the corrupting influences of drugs on people and institutions, and the difficulties faced by Colombian National Police when "orders might begin with great purpose and enthusiasm, but by the time they made their way down the chain of command, with everyone backing away from responsibility, pushing it off on someone else, the great machine of state always wound up confused, impotent, mired," see Mark Bowden, *Killing Pablo: The Hunt for the World's Greatest Outlaw* (New York, NY: Penguin Books, 2002), 122.

18. Rabasa and Chalk, *Colombian Labyrinth*, 12–14.

19. Serafino, "Colombia: Conditions and US Policy Options," 40.

20. Rabasa and Chalk, *Colombian Labyrinth*, 46–47.

21. Audrey K. Cronin, "The 'FTO List' and Congress: Sanctioning Designated Foreign Terrorist Organizations," Congressional Research Service Report, 21 October 2003.

22. Paul E. Saskiewicz, "The Revolutionary Armed Forces of Colombia—People's Army (FARC-EP): Marxist-Leninist Insurgency or Criminal Enterprise?" (Monterey, CA: Naval Postgraduate School Thesis, December 2005), 9; Rabasa and Chalk, *Colombian Labyrinth*, 23–24.

23. Saskiewicz, "The Revolutionary Armed Forces of Colombia," 16–26, 32–35; Rabasa and Chalk, *Colombian Labyrinth*, 24–25.

24. Saskiewicz, "The Revolutionary Armed Forces of Colombia," 27–54; Rabasa and Chalk, *Colombian Labyrinth*, 28–29, 37; Jane's, "Fuerzas Armadas Revolucionarias de Colombia (FARC)," http://www8.janes.com/Search/

documentView.do?docId=/content1/janesdata/binder/jwit/jwit0265.htm@curr
ent&pageSelected=allJanes&backPath=http://search.janes.com/Search&Prod_
Name=JWIT&keyword (accessed 20 November 2008).

25.   Andreas E. Feldmann and Victor J. Hinojosa, "Terrorism in Colombia:
Logic and Sources of a Multidimensional and Ubiquitous Phenomenon,"
Terrorism and Political Violence 21 (2009): 49–50; Rabasa and Chalk, *Colombian
Labyrinth*, 30–34, 45; Jane's, "Ejercito de Liberación Nacional (ELN)," http://
www8.janes.com/Search/documentView.do?docId=/content1/janesdata/binder/
jwit/jwit0266.htm@current&pageSelected=allJanes&backPath=http://search.
janes.com/Search&Prod_Name=JWIT&keyword (accessed 20 April 2009).

26.   Human Rights Watch, *Colombia's Killer Networks: The Military-
Paramilitary Partnership and the United States* (New York, NY: Human Rights
Watch, 1 November 1996), 1–8, http://www.hrw.org/en/reports/1996/11/01/
colombia-s-killer-networks?print (accessed 22 January 2009); Richani, "Caudillos
and the Crisis of the Colombian State," 405–409; Aviles, "Paramilitarism and
Colombia's Low-Intensity Democracy," 385–386; Bowden, *Killing Pablo,* 165–
200.

27.   David Spencer, *Colombia's Paramilitaries: Criminals or Political
Force?* (Carlisle Barracks, PA: Strategic Studies Institute, December 2001), 20.

28.   Spencer, *Colombia's Paramilitaries*, 2.

29.   Aviles, "Paramilitarism and Colombia's Low-Intensity Democracy,"
394–399, 403–407; Spencer, *Colombia's Paramilitaries*, 6–17.

30.   Information collected by the Colombian Commission of Jurists in
Aviles, 403.

31.   Aviles, "Paramilitarism and Colombia's Low-Intensity Democracy," 395–
399; United Nations, "Report by the United Nations High Commissioner for Human
Rights," United Nations Commission on Human Rights, 9 March 1998, http://
www.unhchr.ch/Huridocda/Huridoca.nsf/fb00da486703f751c12565a90059a227/
e84a158767e2e9ecc12566180049cd3e?OpenDocument (accessed 21 April 2009).

32.   Spencer, *Colombian Paramilitaries*, 17–18; Rabasa and Chalk,
Colombian Labyrinth, 32–33.

33.   Manwaring, *Non-State Actors in Colombia*, 75.

34.   See Chapter 7, "Concerning the Public Force," Articles 216–223 in
Republic of Colombia, "Text of the Constitution of Colombia (1991)," http://
confinder.richmond.edu/admin/docs/colombia_const2.pdf (accessed 13 April
2009).

35.   Richard L. Maullin, *Soldiers, Guerrillas and Politics in Colombia*
(Santa Monica, CA: Rand, December 1971), 63.

36.   Thomas C. Bruneau, "Restructuring Colombia's Defense Establishment
to Improve Civilian Control and Military Effectiveness" (Monterey, CA: Center
for Civil-Military Relations, Naval Postgraduate School, 2004), 1, http://www.
resdal.org/experiencias/main-lasa-04.html (accessed 11 March 2009); Notes from
discussions with Colombian and American military and civilian officials.

37.   William C. Spracher, "The Colombian Armed Forces and National
Security," in Zackrison, ed., *Crisis? What Crisis?*

38. Dennis M. Hanratty and Sandra W. Meditz, ed. *Colombia: A Country Study* (Washington, DC: Federal Research Division, Library of Congress, December 1988), 267.

39. Each service chief had the duties of a US military service chief and of an operational commander—transforming his service at the same time as fighting and running it. The ranks of Colombian general officers are brigadier general, major general, and general. The COLMIL commander and the service chiefs hold the rank of general.

40. Farah, "Colombian Rebels Seen Winning War."

41. Manuel Jose Bonett Locarno, *Estrategia General De Las Fuerzas Militares" "Por la seguirdad de la poblacion y sus recurrsos,"* Fuerzas Militares De Colombia Comando General, December 1997; Quotes from Spracher, in Zackrison, *Crisis? What Crisis?*

42. Notes from discussion with a senior Colombian official.

43. Quotes from Spracher, in Zackrison, *Crisis? What Crisis?*, 3.

44. Quote from Myles Frechette, US Ambassador to Colombia from 1994–97, in Larry Rohter, "Armed Forces in Colombia Hoping to Get Fighting Fit," *New York Times*, 5 December 1999, http://proquest.umi.com/pqdweb?index=3&did=46865681&SrchMode=2&sid=1&Fmt=3&VInst=PROD&VType=PQD&RQT=309&VName=PQD&TS=1240516114&clientId=5904 (accessed 5 November 2008).

45. "Colombia's Brass Admit Army Cannot Compete with Guerrilla," *Houston Chronicle*, 25 March 1998, http://proquest.umi.com/pqdweb?index=1&did=27791616&SrchMode=2&sid=1&Fmt=3&VInst=PROD&VType=PQD&RQT=309&VName=PQD&TS=1240519282&clientId=5904 (accessed 5 November 2008).

46. Notes from discussions with Colombian Military officials.

47. Colombia Army, "Disposicion Numero 000011 Por medio de la cual se reorganiza el Ejercito Nacional," 23 August 1996. Document signed by Colombian Army Commanding General Harold Bedoya reorganizing the Colombian Army in accordance with Article 25 of Law 2335 of 1971; comment on US brigade structure, Bruneau, "Restructuring Colombia's Defense Establishment, 2; Information about organizations and incentives below brigade level came from notes from discussions with Colombian and American military officials.

48. Michael Evans, ed., Document 3, US Embassy Cable, 24 May 1988, Part I, "War in Colombia: Guerrillas, Drugs and Human Rights in US-Colombia Policy, 1988–2002," National Security Archive Electronic Briefing Book No. 69, Colombia Documentation Project, http://www.gwu.edu/~nsarchiv/NSAEBB/NSAEBB69/ (accessed 4 August 2008).

49. Evans, Document 3, US Embassy Cable, 24 May 1988, Part I; Human Rights Watch, *State of War: Political Violence and Counterinsurgency in Colombia* (New York, NY: Human Rights Watch, December 1993), 5, http://www.hrw.org/legacy/reports/1993/colombia/statetoc.htm (accessed 26 January 2009); Colombian Army, *Colombian Army Military History* (Bogotá: E3 Section, Army Historical Studies Center, 2007) 284, 389–390.

50. Hanratty and Meditz, *A Country Study*, 283; Colombian Navy, *Closing the Gap: "Towards the Future"—The Naval Strategy, a Cornerstone in the Fight against Narco-terrorism* (Bogotá: Colombian Navy, 2003), 8–27; Ricardo A. Flores, "Improving the US Navy Riverine Capability: Lessons from the Colombian Experience" (Monterey, CA: Naval Postgraduate School Thesis, December 2007), 29–42; Notes from discussions with Colombian Military and civilian officials.

51. US Department of State, "International Narcotics Control Strategy Report, 1997," in Colombia Report, section III, http://www.state.gov/www/global/narcotics_law/1997_narc_report/samer97.html (accessed 1 October 2008).

52. Hanratty and Meditz, *A Country Study*, 264; Notes from discussions with Colombian and American military and civilian officials.

53. US Department of State, "International Narcotics Control Strategy Report, 1997"; Notes from discussions with Colombian and American military and civilian officials.

54. See Articles 213 and 221 in Republic of Colombia, "Text of the Constitution of Colombia (1991)"; Human Rights Watch, *Colombia's Killer Networks*, Chapter V, "Impunity"; Notes from discussions with Colombian and American military officials.

55. For an overview, see Bradley Coleman, *Colombia and the United States: The Making of an Inter-American Alliance, 1939–1960* (Kent, OH: Kent State University Press, 2008); Kenneth Finlayson, "Colombia: A Special Relationship," *Veritas, Journal of Army Special Operations History* 2, no. 4 (2006): 5–7.

56. For an overview, see Bradley Coleman, "The Colombian Army in Korea, 1950–1954," *Journal of Military History* (January 2005): 1137–1177; Charles H. Briscoe, "Barbula and Old Baldy, March 1953: Colombia Heaviest Combat in Korea," *Veritas, Journal of Army Special Operations History* 2, no. 4 (2006): 15–23, 24–29; Charles H. Briscoe, "Across the Pacific to War: The Colombian Navy in Korea, 1951–1955," *Veritas, Journal of Army Special Operations History* 2, no. 4 (2006): 24–29.

57. Charles H. Briscoe, "Colombian *Lancero* School Roots," *Veritas, Journal of Army Special Operations History* 2, no. 4 (2006): 30–37.

58. Dennis M. Rempe, "Guerrillas, Bandits, and Independent Republics: US Counter-insurgency Efforts in Colombia, 1959–1965," *Small Wars and Insurgencies* (Winter 1995): 304–327; Dennis M. Rempe, *The Past as Prologue? A History of US Counterinsurgency Policy in Colombia, 1958–1966* (Carlisle Barracks, PA: Strategic Studies Institute, March 2002); Dennis M. Rempe, "The Origin of Internal Security in Colombia: Part I-A CIA Special Team Surveys *la violencia*, 1959–1960," *Small Wars and Insurgencies* (Winter 1999): 24–61; Charles H. Briscoe, "*Plan Lazo*: Evaluation and Execution," *Veritas, Journal of Army Special Operations History* 2, no. 4 (2006): 38–46.

59. Charles H. Briscoe, "Blue Helmets to Maroon Berets: *Batallon Colombia* in the Suez and Sinai, 1956–1958, 1982–2006," *Veritas, Journal of Army Special Operations History* 2, no. 4 (2006): 103–106.

60. David C. Wolfe, "The View from Washington," in Zackrison, *Crisis? What Crisis?*

61. Evans, "War in Colombia," Part II, "Counterdrug Operations."

62. Tom Brown, "Colombia Official Reveals Shake-Up in Military's Echelon," *Houston Chronicle*, 1 November 1996, http://proquest.umi.com/pq dweb?index=0&did=23117318&SrchMode=2&sid=1&Fmt=3&VInst=PROD &VType=PQD&RQT=309&VName=PQD&TS=1240931257&clientId=5904 (accessed 4 November 2008).

63. James L. Zackrison, "Executive Summary: Workshop on Security Issues in Colombia," in Zackrison, *Crisis? What Crisis?*

64. US General Accounting Office, *Drug Control: US Counternarcotics Efforts in Colombia Face Continuing Challenges* (Washington, DC: Government Printing Office, February 1998), 15.

65. Robert Noriega, "United States Priorities in Colombia," in Zackrison, *Crisis? What Crisis?*

66. Mark Peceny and Michael Durnan, "The FARC's Best Friend: US Antidrug Policies and the Deepening of Colombia's Civil War in the 1990's," *Latin American Politics and Society* (Summer 2006): 95–116.

67. Patricia Bibes, "Colombia: The Military and the Narco-Conflict," *Low Intensity Conflict & Law Enforcement* (Spring 2000): 33.

68. Coletta Youngers, "US Entanglements in Colombia Continue," *NACLA Report on the Americas*, March/April 1998, http://proquest.umi.com/pqdweb ?did=28811023&sid=4&Fmt=3&clientid=5094&RQT=309&VName=PQD (accessed 5 November 2008).

69. See US Department of State, "Colombia Country Report on Human Rights Practices for 1997" (Washington, DC: US DOS), http://www.state.gov/www/global/human_rights/1997_hrp_report/colombia.html (accessed 16 March 2009).

70. See US Department of State, "Colombia Country Report on Human Rights Practices for 1996" (Washington, DC: US DOS), http://www.state.gov/www/global/human_rights/1996_hrp_report/colombia.html (accessed 16 March 2009).

71. Since 1986, section 490 of the Foreign Assistance Act required the President to certify by 1 March each year that countries receiving US counternarcotics assistance were cooperating in meeting US counterdrug objectives. A partial certification for vital national interest for uncooperative countries provided a middle ground of limited support between full support and no support. GAO, *Drug Control: US Counternarcotics Efforts in Colombia Facing Continuing Challenges,* 18–21.

72. Nina M. Serafino, "Colombia: The Problem of Illegal Narcotics and US–Colombian Relations," Congressional Research Service Report, updated 11 May 1998, 8–10.

73. Serafino, "Colombia: The Problem of Illegal Narcotics and US–Colombian Relations," 10.

74. GAO, *Drug Control: US Counternarcotics Efforts in Colombia Facing Continuing Challenges*, 35–41.

75. GAO, *Drug Control: US Counternarcotics Efforts in Colombia Facing Continuing Challenges*, 9–10.

76. Nina M. Serafino, "Colombia: Summary and Tables on US Assistance, FY1989–FY2003," Congressional Research Service Report, 3 May 2002, 6; Foreign Military Sales (FMS) are government-to-government sales of US defense equipment, training, or services. DOS, not DOD, funds the Foreign Military Assistance Program (FMAP), which includes foreign military financing (FMF) of DOD equipment, training, and services through grants or loans and International Military Education and Training (IMET) for training of personnel in the United States. DOS funds anticrime and antidrug programs. DOD provides Section 1004 counterdrug training support (CDTS) when requested by law enforcement for host nation police and military forces involved in counterdrug operations and Section 1033 riverine support when requested for units involved in counterdrug operations. See US Department of State, "Foreign Military Training and DOD Engagement Activities of Interest: Joint Report, Volume 1" (Washington, DC: US DOS, March 2002), "II. Description of Programs," http://www.state.gov/t/pm/rls/rpt/fmtrpt/2002/10607.htm (accessed 1 October 2008).

77. Dean A. Cook, "U.S. Southern Command: General Charles E. Wilhelm and the Shaping of U.S. Military Engagement in Colombia, 1997–2000," in *America's Viceroys: The Military and U.S. Foreign Policy,* ed. Derek S. Reveron (New York, NY: Palgrave Macmillan, 2004), 136–142. In 1999, USARSO moved to Fort Buchanan, PR, and then to Fort Sam Houston, TX, in 2002. In 1999, AFSOUTH or 12th Air Force moved to Davis-Monthan Air Force Base, AR. When USNAVSO moved to Mayport, FL, in 2004, USMARFORSOUTH moved to Miami and USSOCSOUTH moved to Homestead, FL. Marine assets came from Camp Lejeune, NC, and Army Special Forces (SF) teams from 7th SF Group at Fort Bragg, NC.

78. Charles E. Wilhelm, "Statement before the 105th Congress Committee on Armed Services, United States Senate," 5 March 1998, http://www.fas.org/irp/congress/1998_hr/s980305w.htm (accessed 15 October 2008).

79. US General Accounting Office, *School of the Americas: US Military Training for Latin American Countries* (Washington, DC: GAO, August 1996). The school reopened in January 2001 at Fort Benning, GA, as the Western Hemisphere Institute for Security Cooperation (WHINSEC).

80. Christopher W. Muller, "USMILGP Colombia: Transforming Security Cooperation in the Global War On Terrorism" (Monterey, CA: Naval Postgraduate School Thesis, December 2006), 13, 42. Each security assistance organization has similar missions, but an organization and a name or title unique to that country. Muller noted 16 different titles when writing his paper. After September 2001, temporary duty (TDY) personnel increased in number and they served 3-to-6-month TDY tours. The turbulence of constantly rotating personnel created continuity and program challenges for the MILGP.

81. Bowden, *Killing Pablo*, 54.

82.   Geoffrey B. Demarest, "The Overlap of Military and Police in Latin America," April 1995, http://www.smallwars.quantico.usmc.mil/search/Lessons Learned/LatinAm/milpolre.htm (accessed 29 August 2008).

83.   Cook, "General Charles E. Wilhelm and the Shaping of U.S. Military Engagement in Colombia," in Reveron, *America's Viceroys,* 133.

84.   Douglas Porch and Christopher W. Muller. "'Imperial Grunts' Revisited: The US Advisory Mission in Colombia," in *Military Advising and Assistance: From Mercenaries to Privatization, 1815–2007*, ed. Donald Stoker (London: Routledge, 2008), 170.

85.   US General Accounting Office, *Military Training: Management and Oversight of Joint Combined Exchange Training* (Washington, DC: GAO, July 1999), 7.

86.   GAO, *Military Training,* 47–48; Dana Priest, "US Force Training Troops in Colombia; Exercise Anti-Drug Efforts," *Washington Post*, 25 May 1998, http://proquest.umi.com/pqdweb?index=5&did= 29679864&SrchMode=2&sid=1&Fmt=3&VInst=PROD&VType=PQD&RQT=309&VName=PQD&TS=1241189705&clientId=5904 (accessed 5 November 2008).

87.   "Colombian Guerrillas Ambush Army Convoy, Kill 31 Soldiers," *San Antonio Express–News,* 17 April 1996, http://proquest.umi.com/pqdweb?index=8&did=15244774&SrchMode=2&sid=2&Fmt=3&VInst=PROD&VType=PQD&RQT=309&VName=PQD&TS=1241197937&clientId=5904 (accessed 17 November 2008); Colombian Army, *Colombian Army Military History*, 333.

88.   Quoted in Thomas A. Marks, "Colombian Army Counterinsurgency," *Crime, Law & Social Change* (July 2003): 84–85.

89.   Colombian Army, *Colombian Army Military History*, 333; Christopher Torchia, "Weakness of Army Prolongs Colombian War," *Colombian*, 2 September 1996, http://proquest.umi.com/pqdweb?index=6&did=15312706&SrchMode=2&sid=1&Fmt=3&VInst=PROD&VType=PQD&RQT=309&VName=PQD&TS=1241202373&clientId=5904 (accessed 5 November 2008); David Spencer, "Latin America, A Lesson for Colombia," *Jane's Intelligence Review,* 1 October 1997, http://search.janes.com/Search/documentView.do?docId=/content1/janesdata/mags/jir/history/jir97/jir00125.htm@current&pageSelected=allJanes&keyword=david%20spencer&backPath=http://search.janes.com/Search&Prod_Name=JIR& (accessed 4 March 2009).

90.   Gabriella Gamini, "Colombia Fights Back as Guerrilla Launch Fiercest Raids in Decades," *Times*, 3 September 1996, http://proquest.umi.com/pqdweb?did=34285574&sid=5&Fmt=3&clientId=5094&RQT=309&VName=PQD (accessed 5 November 2008).

91.   Jose de Cordoba, "Guerrilla Attacks Kill 100 in Colombia in Possible Link to Nation's Drug War," *Wall Street Journal,* 3 September 1996, http://proquest.umi.com/pqdweb?index=10&did=23868034&SrchMode=2&sid=2&Fmt=3&VInst=PROD&VType=PQD&RQT=309&VName=PQD&TS=1241204004&clientId=5904 (accessed 5 November 2008).

92.   Georgie Geyer, "For Colombia, 1996 Nothing But Trouble," *Tulsa World,* 31 January 1997, http://proquest.umi.com/pqdweb?index=192&did=168

24144&SrchMode=1&sid=7&Fmt=3&VInst=PROD&VType=PQD&RQT=309 &VName=PQD&TS=1239811252&clientId=5904 (accessed 5 November 2008).

93. Yadira Ferrer, "Change of Commander in Chief a Boost to Peace," *Inter Press Service*, 25 July 1997, http://proquest.umi.com/pqdweb?did=13182995&s id=4&Fmt=3&clientId=5094&RQT=309&VName=PQD (accessed 5 November 2008).

94. "Another Beating," *The Economist,* 10 January 1998, http://proquest. umi.com/pqdweb?did=25330049&sid=4&Fmt=3&clientid=5094&RQT=309 &VName=PQD (accessed 5 November 2008); David Spencer, "Focus—Latin America, Bogotá Continues to Bleed as FARC Find Their Military Feet," *Jane's Intelligence Review,* 1 November 1998, http://www4.janes.com/subscribe/jir/doc_ view.jsp?K2DocKey=/content1/janesdata/mags/jir/history/jir98/jir00815.htm@ current&Prod_Name=JIR&QueryText=%3CAND%3E%28%3COR%3E%28%2 8%5B80%5D%28+colombia+%3CAND%3E+david+%3CAND%3E+spencer% 29+%3CIN%3E+body%29%2C+%28%5B100%5D+%28%5B100%5D%28+co lombia+%3CAND%3E+david+%3CAND%3E+spencer%29+%3CIN%3E+title %29+%3CAND%3E+%28%5B100%5D%28+colombia+%3CAND%3E+david +%3CAND%3E+spencer%29+%3CIN%3E+body%29%29%29%29 (accessed 4 March 2009).

95. Spencer, "Latin America, A Lesson for Colombia."

96. Diana Schemo, "US Is to Help Army in Colombia Fight Drugs but Skeptics Abound," *New York Times,* 25 October 1997, http://proquest.umi.com/ pqdweb?index=23&did=20044995&SrchMode=2&sid=4&Fmt=3&VInst=PRO D&VType=PQD&RQT=309&VName=PQD&TS=1239807437&clientId=5904 (accessed 5 November 2008).

97. Cronin, "The 'FTO List' and Congress."

98. Farah, "Colombian Rebels Seen Winning War."

99. Spencer, "Focus—Latin America, Bogotá Continues to Bleed as FARC Find Their Military Feet"; Rabasa and Chalk, *Colombian Labyrinth*, 42.

100. Brooks, "Colombian Military Is Called To Account: Rebels Outwitted Forces, Critics Say."

101. "Colombia's Brass Admit Army Cannot Compete with Guerrilla"; Notes from discussions with Colombian and American military officials confirm a detailed Colombian Army investigation later revealed a major institutional and situational weakness at El Billar.

102. Charles E. Wilhelm testimony in US House of Representatives. *Transcript, Hearing of the House Committee on International Relations on US Anti-Drug Policy Towards Colombia*, 31 March 1998, http://www.colombiasupport. net/199803/hr033198.html (accessed 15 October 2008).

103. Rosso José Serrano testimony in US House of Representatives. *Transcript, Hearing of the House Committee on International Relations on US Anti-Drug Policy Towards Colombia*, 31 March 1998. http://www.colombiasupport. net/199803/hr033198.html (accessed 15 October 2008).

104. Farah, "Colombian Rebels Seen Winning War."

105. Ben Barber, "Colombia Needs Elections, Arms Aid, Ex-Army Chief Says," *Washington Times,* 22 March 1998, http://proquest.umi.com/pqdweb?index=209&did=27643577&SrchMode=1&sid=1&Fmt=3&VInst=PROD&VType=PQD&RQT=309&VName=PQD&TS=1239818287&clientId=5904 (accessed 5 November 2008).

106. Notes from discussions with American military officials; "Colombia Targets Rebel Stronghold: 5,000-Strong Unit Will Patrol Southern Jungle," *Rocky Mountain News*, 13 March 1998, http://proquest.umi.com/pqdweb?did=28564227&sid=3&Fmt=3&clientld=5094&RQT=309&VName=PQD (accessed 5 November 2008).

107. Laura Brooks. "Colombia Disbands Controversial Army Brigade," *Washington Post*, 21 May 1998, http://proquest.umi.com/pqdweb?index=119&did=29617803&SrchMode=1&sid=1&Fmt=3&VInst=PROD&VType=PQD&RQT=309&VName=PQD&TS=1239820613&clientId=5904 (accessed 4 November 2008).

108. Kristian D. Skinner, "An Historical Analysis of the Colombian Dilemma" (Washington, DC: National Defense University Student Paper, April 2001), 28–29.

109. Spencer, "Focus—Latin America, Bogotá Continues to Bleed as FARC Find Their Military Feet"; Colombian Army, *Colombian Army Military History,* 334, 341–342.

Chapter 2

"Change Begins Today"[1]

The Pastrana Presidency: Negotiations with the Guerrillas and Plan Colombia (1998–2002)

From the times of General Fernando Tapias the Military Forces were restructured and its mentality changed, and we undertook a plan in which every day we are performing in an increasingly more professional and technical manner and in keeping with the Constitution and the law.

General Freddy Padilla, COLMIL Commander[2]

We are committed to maintaining the line between counterinsurgency and counterdrugs, because we are not in the counterinsurgency business.

US Official[3]

Forty-three-year-old Andres Pastrana became Colombia's 60th President on 7 August 1998 following a week of widespread Revolutionary Armed Forces of Colombia (FARC) and National Liberation Army (ELN) attacks that caused, among other things, the destruction of an important Colombian National Police (CNP) antinarcotics base at Miraflores. Elected on a peace platform and promising to pursue a settlement with the guerrillas, Pastrana had initiated talks with the FARC before his inauguration. The son of a former president, as well as a former senator and mayor of Bogotá, Pastrana promised Colombians "change begins today" to indicate a break with the corruption-ridden Samper administration. However, Pastrana inherited serious problems that complicated his job: deteriorating security situation, serious economic crisis, and lack of a legislative majority to support his peace platform. In addition, after the recent attacks questions remained about the willingness of the guerrillas to negotiate in good faith, the capabilities of the security forces to maintain the status quo, and the potential changes in the relationship with the United States.[4]

President Pastrana Begins Negotiations: 1998

In creating his new security team, Pastrana appointed Rodrigo Lloreda as Minister of Defense and kept only one incumbent—Major General Rosso Jose Serrano, the US-supported, counterdrug-supporting commander of the CNP. Although some in the American press referred to these changes as a purge given the guerrilla attacks, but at the beginning of each presidency and 2 years afterward, this change normally occurred. Pastrana chose General Fernando Tapias—the Colombian Army deputy, a former Army inspector general, and a former mobile infantry brigade (BRIM) commander—to be his Armed Forces commander. General Jorge Enrique Mora, commander of the Fifth Division near Bogotá and a former BRIM commander became the Army commander. The deputies of the Colombian Navy and the Air Force—Admiral Sergio Edilberto Garcia and General Jose Manuel Sandoval—each rose to become service chiefs. These commanders would attempt to revitalize the security forces while dealing, in Minister Lloreda's words, "with a delicate security situation and the reconciliation process that are government priorities."[5] This proved difficult. In mid-August, a 600-man FARC unit mauled a counterguerrilla battalion and a regular infantry battalion from the 17th Brigade in a bitter 3-day fight in northwest Colombia—a violent drug and arms smuggling area just south of Panama. Reports of over 40 killed and 130 missing brought both Tapias and Mora to the scene.[6] Making improvements while fighting with inadequate and dispersed forces presented a serious military challenge. Unlike other presidents who had negotiated with the guerrillas, Pastrana did not reduce the Ministry of National Defense (MOD) budget. Although the military opposed the Pastrana approach, it had to assess its weaknesses and identify requirements within its limited resources. Overcoming a defensive, defeatist mindset—one that made the guerrillas appear stronger than they really were—proved the initial challenge. Although the defeats at El Billar and Miraflores had served as a "catalyst to learn" for the military forces, relations with the CNP remained strained. Other than reacting to guerrilla attacks and withdrawing isolated security forces from threatened areas, little had been done after the August attacks and the destruction of the Miraflores base, which would not be permanently occupied by security forces for another 5 years.[7] This would change on 1 November.

Before agreeing to negotiations with the Pastrana government, the FARC made five demands:

(1) The government would give the FARC control of five municipalities—four in Meta and one in Caqueta departments.

(2) The government security forces would fight the *autodefensas*.

(3) The government would permit social protests and public demonstrations.

(4) The government would develop an alternate crop substitution and economic development plan.

(5) The government would stop its wanted poster and financial incentives program directed against the FARC leadership.[8]

Each demand undercut government programs, dispersed scarce government resources, and worked to establish the FARC as a legitimate political organization with legal or belligerent status. Coming to the negotiations with a position of increasing strength—political, psychological, and military, the FARC intended to increase its power by extending the negotiations over time by pursuing a make-war-and-talk-peace policy. In mid-October, Pastrana agreed to withdraw Colombian security forces from the municipalities of Mesetas, La Uribe, Vista Hermosa, San Vincente de Caguan, and La Macarena by 7 November for a period of 90 days. This provided the FARC temporary control over a 42,000 square kilometer *zona de despeje*, or cleared zone, with over 90,000 inhabitants. Established in southeastern Colombia in a region long under FARC influence, the *zona* had a land area equal to Switzerland or two El Salvadors or Maryland and Delaware combined. (See Figure 7, *Zona de despeje*.) The government of Colombia and the FARC agreed to meet on 7 January 1999 to begin negotiations. At the time, no one expected the *zona* would be extended 11 times in the following 38 months.[9] The result would be a sanctuary in which the FARC would establish control, increase its military strength and prepare military attacks, hold and ransom hostages, grow and process illegal drugs, smuggle illegal drugs and arms, and establish ties with such diverse groups as Latin American and European Union politicians, nongovernmental organizations (NGOs), humanitarian organizations, drug traffickers, and other guerrilla and terrorist groups. During the whole time, the guerrillas continued to conduct military operations throughout the country.

On the eve of the establishment of the *zona*, the FARC attacked Mitú—an isolated town of 15,000 inhabitants similar in size to Miraflores but the capital of Vaupes department in southeastern Colombian near the Brazilian border. Supported by almost 400 propane gas cylinder bombs, over 1,000 FARC guerrillas flattened the police station, destroyed the communications tower, and captured the local airstrip as they overran the 125-man police garrison after 12 hours of fighting. They took about 45 prisoners—adding to the almost 250 police and military captives held by the FARC. To some it appeared the FARC had captured its first department capital. The Colombia Military (COLMIL) reacted immediately. Reachable only by air or river and with no airstrip in the jungle near Mitú, the Colombian Air Force secured an airstrip inside Brazil from which to refuel the transport aircraft delivering combat troops and the helicopters to ferry them into battle. After a 3-day battle, a force of over

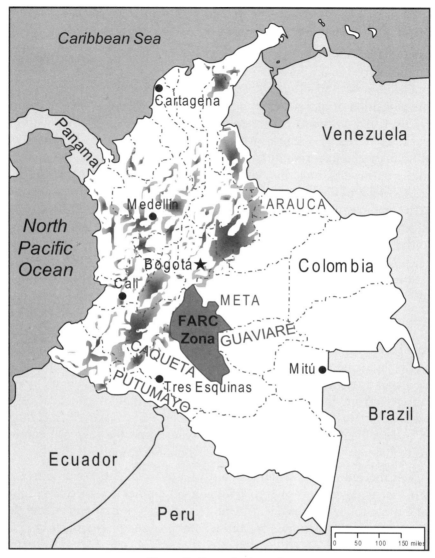

Figure 7. *Zona de despeje.*

400 Colombian Army soldiers—including 52 Counterguerrilla Battalion (BCG) rebuilt after El Billar and soldiers from the 7th Brigade—and CNP personnel ejected the FARC from Mitú.[10] The COLMIL acted quickly and effectively, but concerns arose about a lack of preparedness against a known impending attack. A failure to share information meant that the CNP had reinforced its garrison before the attack and the Colombian Army had failed to act until after the attack. Other problems arose about the same time. In mid-November, General Hector Fabio Velasco became Air

Force commander when his predecessor resigned "as a point of honor," according to a Department of State (DOS) report, "setting a laudable precedent for acceptance of responsibility" after a C-130 aircraft carrying illegal drugs landed in Miami.[11] Publically accepting full responsibility for the performance of the Armed Forces, General Tapias promised:

> . . . to fulfill my responsibilities as commander and deliver a trained and motivated unit, a unit with defense capabilities, within 90 days at the latest. . . . This commitment means that I will accept responsibility for, in advance, the consequences of the actions or omissions and accept with character, frankness and moral courage any responsibility for failures or setback that are blameworthy and result from carelessness, neglect or abandonment of duties, or the undue execution of orders received or the missions assigned.[12]

Although each of the services required reorientation, Tapias stated that the Army would be reorganized starting "with the institution's own professional core to face the challenge posed by the subversive groups."[13] Reorganization and reform would take time, additional resources, hard work, and dedicated leadership, but after Mitú the Army, under General Mora, had committed itself to addressing its deficiencies. How long this would take and how well it would be done remained unknown.

After the inauguration of President Pastrana, the United States sought a closer, more cooperative relationship with Colombia. Ambassador Curtis Kamman supported the peace process, which included talks with the ELN in October, continued judicial reform, and improvements in counterdrug efforts. In its annual human rights report, the DOS noted some human rights improvements—the inactivation of the notorious 20th Intelligence Brigade in May, the decision to disband most Community Associations of Rural Vigilance (CONVIVIR) units in July, the repeated statements of Pastrana and Tapias that ties between military personnel and the *autodefensas* would not be tolerated—but overall the Colombian Government record remained poor.[14] At the same time, the commander of the United States Southern Command (USSOUTHCOM) continued his military-to-military visits. In 1998, 12 US Special Forces (SF) mobile training teams (MTT) from the 7th SF Group conducted short, counterdrug training courses for a small number of COLMIL and CNP personnel. A similar number of Navy special operations and Marine MTTs worked with the Colombian Navy riverine units. Total US military personnel in Colombia had remained below 200.[15] In fiscal year 1998, the CNP received about 90 percent of the US counterdrug assistance, while the COLMIL—basically the Air Force and Navy given the Army's resistance to human rights vetting—received the remainder to maintain radar sites for the Air Bridge Denial Program and to expand riverine units. At a 2 December meeting in Colombia between the Minister of Defense and the US Secretary of Defense, the United States decided to "cautiously reengage" the COLMIL. An American official reemphasized the US commitment "to maintain the line between counterinsurgency and counterdrugs, because we are not in the counterinsurgency business." After the meeting, Minister Lloreda stated that a Colombian–US military task force would address "the modernization of the Colombian Military to restructure [and] . . . to focus on its mobility, its sustainability, its intelligence capabilities, its command and control." In addition, the United States would assist in strengthening the skeletal Joint Task Force–South (JTF-S) at Tres Esquinas in southern Colombia by training and partially equipping by mid-1999 a 1,000-man Army counterdrug brigade to work with the Antinarcotics Police Directorate (DIRAN), setting up a joint police-military intelligence center to support those counterdrug operations, and establishing an electronic listening post to gather intelligence.[16] By the end of 1998, the COLMIL, particularly the Army, had committed itself to reorganizing to fight the guerrillas; the United States had committed itself to supporting counterdrug efforts through the Narcotics Affairs Section (NAS) and the Military Group (MILGP); and President Pastrana had committed himself to negotiations.

Laying a Foundation: 1999

During the first half of 1999, the MOD took the initial steps in what would be a multiyear trial-and-error process to transform the Armed Forces through institutional changes, new technologies, and new doctrine to address the internal security threats. The prognosis was uncertain, the results unknown. Many believed this would require "at least a generation to really turn the Armed Forces around." An analysis of its 1997–98 experiences and the brutally frank report on El Billar provided the COLMIL leadership insights into its shortcomings and deficiencies that would require immediate action, fundamental change, and long-term programs to address. First, Tapias told the Armed Forces, "We are at war, and all our resources must be dedicated to operations."[17] Then, the Colombian Army leadership reinforced these words by actions as units moved from garrison into the field. Second, to increase the number of troops available for combat, the Army began replacing conscripts exempt from combat with professionals. By 2001, this professionalization provided an additional 30,000 combat troops without increasing the Army end strength. Third, the Colombian Army retrained its units on counterguerrilla skills and night operations. In 1998, 90 percent of the Colombian Army training had been conventional and 10 percent counterguerrilla; in 1999, these percentages began to reverse. Furthermore, the Colombian Air Force began to focus more on close air support for military and police units in combat and on night operations. Through this process—coupled with military successes—morale improved as an offensive, aggressive mindset replaced a defensive, defeatist one. In addition to moving from a regular to a professional force and to training for counterguerrilla war, other important long-term transformation areas included developing a military strategy, joint operations, mobility—helicopters and aircraft, intelligence capabilities, training and education systems, reorganized headquarters, new types of units, riverine capabilities, and human rights training.[18] Although informed by US military concepts and suggestions, this transformation was Colombian-initiated, Colombian-led, and counterguerrilla-focused.

At the same time, the COLMIL pursued another trial-and-error effort to reengage the US military through a limited counterdrug effort. Gaining access to US counterdrug funds, training, helicopters, and support motivated this effort. First, the COLMIL addressed its Plan Condor, which focused on reducing overall violence. Second, it worked to increase the capabilities of JTF-S by providing new and better-trained units. Third, it began to organize a military counterdrug joint intelligence center (JIC-S) at JTF-S. In March, the United States agreed to increase the sharing of counterdrug intelligence with JIC-S. However, the CNP—not the COLMIL—led the counterdrug effort focused on coca eradication in the Guaviare department. For the COLMIL, this counterdrug program required not only US assistance, but also a closer working relationship with the CNP. Nevertheless, for some Americans "the most important change" proved to be the "introduction of mandatory human rights courses for most military personnel"—not the COLMIL reform effort or its willingness to participate in the counterdrug program. Although understanding the need to improve its human rights record, few in the COLMIL at that time would have equated mandatory classes as "the most important change" to deal with their primary concern—the guerrillas, or to address a secondary effort—the drug problem, or to improve its human rights record.[19]

In March, President William Clinton certified Colombia for counterdrug funding, which lifted previous military assistance restrictions. However, Congress attached the Leahy amendment to the defense budget, which made human rights vetting by DOS a requirement for any military assistance for Colombia, not just counterdrug assistance.[20] That same month, in

Congressional testimony General Charles E. Wilhelm praised the Pastrana civil-military team's cooperation and commonsense approach. He stated that for "the first time in years . . . [he was] confident that Colombia's military leadership is equal to the task at hand." Acknowledging human rights as "Colombia's international Achilles heel," he added, "I am personally convinced that there are no institutional linkages between the paramilitaries and the Army, but I cannot rule out local collusion." In his review of ongoing USSOUTHCOM programs for Colombia, Wilhelm mentioned assisting in the reform and restructure of the COLMIL, training a Colombian Army counterdrug battalion to work with the CNP, and conducting a 5-year riverine program in Colombia and Peru.[21]

Putting together a counterdrug package for JTF-S at Tres Esquinas proved challenging. With no programmed funding, Wilhelm obtained $7 million from the Department of Defense (DOD) counterdrug funds to train and equip a counterdrug battalion and to establish a joint intelligence center–south (JIC-S) and a joint operations center–south (JOC-S) and another $5 million for infrastructure. He considered the 1,000-man counterdrug battalion—modeled on a US Ranger battalion and larger than three Colombian BRIMs—to be a pilot program for working with the Colombian Army.[22] When a DOS counterdrug official offered 14 UH-1N helicopters for shipping costs to the COLMIL, the Air Force declined for maintenance and cost reasons. The Army accepted the UN-1N helicopters and its aviation battalion—four or five MI-17 helicopters—grew in size by 300 percent.[23] In April, the first of several US SF training teams rotated into Colombia to begin training human rights vetted personnel from which the counterdrug battalion would be formed. Despite efforts to improve Colombian–US military relations, things became "more conditional and tenuous" in April when President Pastrana, under heavy American pressure, relieved three general officers—the Colombian Army director of operations, a division commander, and the war college director—for alleged past ties to the illegal *autodefensas*. After that, for the COLMIL—and particularly the Army—"the United States went from the cavalry riding to the rescue to just another element in the ongoing crisis, a generally positive force but one which would have to be assessed critically and watched."[24]

Although the initial January talks with the FARC had stalled almost immediately, President Pastrana continued to work to find common ground for negotiations with the guerrillas. When the 90-day limit for the *zona de despeje* expired in February, he extended it for another 90 days despite continued FARC attacks, kidnappings, and vows never to lay down its arms. In late May, the announcement that the *zona*—or what some called Farclandia—would be extended indefinitely as long as peace talks continued brought an immediate reaction from the MOD. Minister Lloreda resigned, followed by many military officers, stating, "Too many concessions have been made by the government, and that's the perception of the overwhelming majority of Colombians." In fact, almost 80 percent of Colombians opposed the *zona*. The FARC, he added, "are preparing more for war than for peace." Lloreda stressed that to make such strategic decisions about the *zona* without consulting both the MOD and the COLMIL beforehand—in this case, neither had been consulted— was "incalculably dangerous." In response to the unprecedented mass resignation of general officers and hundreds of military officers, Pastrana met with his military commanders.[25] The resignations indicated that the COLMIL "senior leadership did not believe they were wrong, or that their methods unsuccessful; they believed the President was reckless in making concessions to the armed left."[26] Accepting only the resignation of the Minister of Defense, Pastrana replaced him with Luis Fernando Ramirez. Thereafter, coordination of peace negotiations and security issues improved somewhat.

Early in 1999, the FARC warned of a "first great offensive" if peace negotiations failed to progress. Although the government made numerous concessions, the FARC made none and refused to accept a ceasefire. Public support fell to where 75 percent of Colombians polled considered the FARC a terrorist group with no ideology except kidnapping and drug trafficking. On the eve of the talks scheduled for 7 July, the FARC postponed them for 2 weeks. Then within days, the FARC launched its largest offensive in 40 years. Striking primarily in multiple directions from the *zona de despeje*, FARC forces from its Bloque Oriental—most supported by propane gas-cylinder bombs and some by homemade armored tractors— attacked military and police installations in 10 departments—primarily in eastern and southern Colombia. One group attempted to isolate Bogotá by blocking the roads to the capital. Unlike previous offensives, Colombian Army forces held their own and within a week or so of serious fighting, the Fourth Division—commanded by Brigadier General Carlos A. Ospina, a former BRIM commander—blunted the attacks and forced the FARC back into its safe haven in the *zona*. "This offensive," according to General Tapias, intended "to show an image of strength before the opening of peace negotiations with the national Government." Unlike previous attacks, the FARC strength met effective Army strength reinforced by Air Force support, but neither side dominated the other. A conference of Catholic bishops issued a pastoral letter that forecast "a dark panorama that has no limits" in which "clear possibilities of success cannot be seen, and the civilian population is less and less secure in a country buffeted by the winds of war and death."[27]

Like the guerrillas, the United Self-Defense Groups of Colombia (AUC)—another active player in this "dark panorama"—grew in strength during the year from 5,000 to 7,000 combatants organized in seven major groups. The DOS reported that many *autodefensa* members had security force or guerrilla backgrounds and that these illegal groups varied from those arising from the legitimate desire of rural Colombians to defend themselves against the guerrillas to paid private armies working for prominent landowners or drug traffickers. Their campaign of terror and intimidation focused on guerrilla supporters rather than the armed groups. This increased internal displacements as Colombians migrated from violence-prone areas to the cities. The number of civilian massacres—defined as the killing of three or more persons at the same place—reached almost 400 with over 1,800 dead.[28] The *autodefensas* committed the majority of the killings, and the guerrillas had the next highest number. Human Rights Watch reported that half of the 18 Army brigades had links to paramilitary activity. It charged that the brigade in Cali had created the paramilitary Calima Front to attack the ELN after the May kidnapping of 140 worshipers from a church. Observing that "training alone, even when it includes human rights instruction, does not prevent human rights abuses," the Human Rights Watch called on the Colombian Government to take action.[29] Several years before, the COLMIL had initiated human rights training that included role-playing exercises, classroom instruction, and written examinations. By 1999, about 90 percent of the military and all of the CNP had received training. Even though allegations of human rights abuses by the military dropped from 2,000 in 1996 to just over 300 in 1998 to only 40 by August 1999, human rights groups continued to attack the COLMIL for alleged links to the paramilitaries.[30] Earlier in the year, Pastrana had relieved three general officers being investigated for ties to the *autodefensas*. On 12 August, he signed a revised Military Penal Code with stronger measures and approved the creation of an independent military lawyer or Judge Advocate General Corps. Despite repeating that collusion between military personnel and the paramilitaries would not be tolerated and that the Colombian Army would combat the *autodefensas*, "security force actions in the field were not always consistent with the leadership's positions." Meeting the FARC and ELN guerrilla threat stretched an overextended Army. A senior Army officer suggested that the CNP should lead the effort against the *autodefensas*.[31] Viewed by many Colombians as antiguerrilla rather than antigovernment, the illegal paramilitaries—regardless of origin—would not become a security force priority for years.

After the July attacks, the COLMIL continued its primary focus on making the military forces more effective against the guerrillas by better training, intelligence, air support, and leadership. At the same time, the MILGP concentrated on counterdrug programs. In late July, a US Army RC-7 surveillance aircraft gathering counterdrug intelligence crashed, killing the five Americans and two Colombians on board. At that time, 160 military personnel—including a 12-man SF MTT training Colombian Army counterdrug battalion personnel—and 30 DOD civilians served in Colombia.[32] On 4 August, the Colombian Navy established an 8,000-man, 5-battalion riverine brigade—under the command of its marine infantry brigade—to control almost 8,000 kilometers of rivers throughout the country. Two battalions operated in eastern Colombia, two on rivers flowing to the Caribbean in the north, and one in the southern Pacific region. Each battalion operated in four-boat riverine combat elements (RCE). Each RCE had three 22-foot boats equipped with a .50-caliber machinegun, four M-60 machineguns, and a grenade launcher, and a 26-foot command boat equipped with a .50-caliber machinegun, two M-60 machineguns, an MK-19 grenade launcher, and a 60-mm mortar. Each boat carried 1 officer, 5 noncommissioned officers, and 17 marine infantry professionals. In addition, a counterguerrilla group of 1 officer, 3 noncommissioned officers, and 18 marine infantry served as a strike force. The commander, Admiral Sergio Garcia, said, "If the army and the air force are controlling guerrilla actions, we cannot do less. For this reason we created this brigade . . . to contribute to military actions and to control trafficking in the drugs and chemical precursors that move along Colombia's rivers."[33] This newly established 8,000-man brigade constituted a major component of the Colombian Navy.

Training of counterdrug battalion personnel continued at the Colombian Army's major training base at Tolemaida. Minister of Defense Ramirez commented, "This is the army of the future in Colombia. . . . This is the army we want, and we are going to continue training battalions like this all over Colombia until we have what we need."[34] Given that the Colombian Army had no battalions with 900+ soldiers, Ramirez meant that he wanted training like this—not battalions like this. The counterdrug battalion would support the CNP attack on FARC finances through its coca eradication program. Additional training followed formal activation of the battalion in September. In December, the battalion joined JTF-S—supported by the newly activated JIC-S but not by the helicopters that had been delayed—at Tres Esquinas. The improved mobility sought by the Colombians—primarily through helicopters, but also trucks—failed to arrive as scheduled. The case of 18 Vietnam-era 2.5-ton trucks offered insight into Colombian–US interaction. After waiting for months, the trucks (part of a 30-truck package) arrived in such a dilapidated condition—rusted bodies, batteries and engines so old that the COLMIL had ceased using them a decade earlier, heaters and ignition systems suitable for subzero conditions—that the Colombians, who estimated it would cost roughly $53,000 to make each truck serviceable versus just over $67,000 to buy a new one, refused to accept them. A COLMIL official said, "The only thing they have in common with what was asked for is the tires. What they sent causes more problems than they solve." The United States refused to take the trucks back and left them in Colombia for spare parts. An American official called the Colombian complaints exaggerated, noting that while not meeting all their requirements, the trucks had some useful life—at least useful for the Colombians. A Colombian official offered a different perspective when he said that while officials in Washington took pride in sending aid, "in reality they are sending us millions of dollars in junk."[35] Looking back years later, a senior COLMIL officer observed, "US support was very important for morale," given the isolation and professional embarrassment of decertification, "and not so much for military assistance." Counterdrug assistance, even when serviceable, provided marginal utility in the counterguerrilla struggle.[36]

In November and again in December, the FARC launched multiple attacks from the *zona de despeje* and in northern Colombia. Both times the Colombian Army responded effectively—many units fighting under difficult conditions. However, FARC successes remained few and short lived. When General Tapias announced he had evidence that at least five FARC fronts and mobile columns staged attacks from the *zona*, the government peace commissioner downplayed these charges. On 19 November, the COLMIL commander and his Army commander, General Mora, presented their resignations to President Pastrana. Rejecting the resignations, Pastrana publically defended the military, opposed the exchange of jailed guerrillas for soldier and police hostages, attacked the guerrillas, and served notice that he would not allow the *zona* to turn into a sanctuary from which attacks against the military and the police could be launched. In response, Raul Reyes, a FARC spokesman, threatened to dissolve the *zona* and abandon the peace process.[37] The give and take of negotiations continued, but the FARC and Pastrana both suffered from their failure to reach some agreement. Despite its increasing military strength, the FARC began to lose support for the negotiation process and Pastrana's approval rating dropped to 22 percent by the end of 1999.[38]

Through the COLMIL transformation process, each military service—the 120,000-soldier Army, the 15,000-man Navy, and the 7,500-airman Air Force—ended the year somewhat better prepared to address the immediate counterguerrilla threat and planning to make additional improvements in 2000. Better training, intelligence, and service cooperation had enhanced combat performance, but the military forces remained too small to fight the guerrillas and to secure critical infrastructure throughout the country.[39] A Colombian Army officer captured the frustration of fighting a war under peacetime constraints when he said: "Yet the crucial question is how to control the ground. In our system *everything* is prohibited. If you even attempt to uncover the infrastructure, much less dominate areas, you are violating something. We are in a position of fighting for a system unwilling to defend itself."[40] Despite these challenges, on 23 November the Colombian Army published its reorganization plan for 2000. Each of its five divisions—two with four brigades and three with three brigades—now had nonstandardized brigades that reflected both the limits of force structure and the task organization for specific areas. The Colombian Army divisions totaled 45 infantry battalions, 9 cavalry squadrons, 7 artillery battalions, 11 engineer battalions, 5 military police battalions, 45 counterguerrilla battalions, 17 support battalions, and 16 new antikidnapping units (for example, GAULA). The reorganization included a three-battalion counterdrug brigade and a rapid reaction force (*Fuerza de Desplique Rapido* or FUDRA) composed of the three BRIM, the four-battalion SF brigade—more ranger battalions than US SF units—and the Colombian Army aviation brigade.[41] Security force helicopters—a scarce and critical mobility asset—numbered a handful for the Colombian Navy, 17 for the Colombian Army, 50 for the Colombian Air Force, and almost 100 for the CNP—half of which were restricted by US assistance to counterdrug operations.[42] Activated in early December, the 5,000-man FUDRA—a small division-equivalent combining the mass of large elite units with the mobility of helicopters—provided the Army a stronger and more flexible response force for countering FARC and ELN attacks. As it looked into 2000, the COLMIL expected additional resources from President Pastrana's initiative known as Plan Colombia.

Plan Colombia: 1999–2000

Many believe that Plan Colombia evolved from the reengagement of the COLMIL by the US military. Some would have it begin with the disbandment of the 20th Intelligence Brigade, the creation of the Army

Military Intelligence Center (CIME), and the establishment of JTF-S in 1998. Others would tie it to the process that birthed the counterdrug battalion, the planning for a counterdrug brigade, and the working relationship between Wilhelm and Tapias. From the beginning, Americans—and later the COLMIL—stressed the need for an overall, comprehensive national plan. A senior Pastrana advisor, Dr. Jamie Ruiz, developed a Colombian concept that included US input. In May 1999, the government issued a white paper titled "Plan Colombia" to address its problems. Colombian– US discussions followed as Wilhelm encouraged Embassy involvement and support.[43] In September, President Pastrana announced his 6-year Plan Colombia. Shortly thereafter, the Colombians and the Americans refined its details. The Colombian Government planned to fund $4 billion of the $7.5 billion cost for the first 3 years and sought international contributions for the remaining $3.5 billion. The United States agreed to fund $1.3 billion, almost 75 percent to address the counterdrug problem, with other contributors providing the remaining $2.2 billion. Because of difficulties raising funds within Colombia and lukewarm international support, only the United States met its funding goal in the first 3 years—creating unintended consequences.[44]

"Plan Colombia: Plan for Peace, Prosperity, and the Strengthening of the State" proved more a concept than a detailed plan. After stating its fundamental goal: "to strengthen the State in order to regain the citizens' confidence and recuperate the basic norms of peaceful coexistence," the plan acknowledged that it would take years to consolidate control throughout the country and to build peace. It briefly described 10 elements or strategies in the following order: economic, fiscal and financial, military, judicial and human rights, counternarcotics, alternate development, social participation, human development, peace, and international. The plan ended by addressing five topics in some detail: economy, counterdrug strategy, justice reform, democratization and social development, and peace process. The only reference to a military strategy—other than discussion in the section on counterdrugs—focused on restructuring and modernizing the security forces "to reestablish the rule of law and provide security throughout the country, and in combating organized crime and armed groups."[45] With Plan Colombia, Pastrana tried to build a peace effort on three components: peace talks with the guerrillas; strengthened security forces; and international assistance in funding economic, political, social, and military programs.[46] Written to address multiple audiences with different goals, Plan Colombia became a nation-strengthening plan in Colombia, a peace plan in Europe, and a counterdrug plan in the United States.

Written with the Americans in mind, the counterdrug strategy section contained the most details and mixed counterdrug with counterguerrilla issues. Acknowledging counterdrug as "one of its top strategic priorities," four "violence-generating agents" were identified: narcotraffickers, guerrillas or subversives, illegal *autodefensas* or paramilitaries, and common criminals. The "National Mission" was "to ensure order, stability, and the rule of law; guarantee sovereignty over national territory; protect the State and the civilian population from threats posed by illegal armed groups and criminal organizations; break the links between the illegal armed groups and the criminal drug industry that supports them." Reduction of cultivation, processing, and distribution of illegal drugs by 50 percent in 6 years became a measurable goal. To reach this goal, six objectives—each with areas of focus—followed:

(1) Strengthening the counterdrug fight through an integrated effort by the Armed Forces to establish military control of areas.

(2) Strengthening the judicial system and fighting corruption.

(3) Neutralizing the drug financial system and seizing its resources.

(4) Neutralizing and fighting agents of violence allied with the drugs.

(5) Integrating national initiatives with regional and international efforts.

(6) Strengthening and expanding alternate development in drug growing areas.

A three-phased plan to reach the 50 percent reduction goal integrated the COLMIL and the CNP efforts to destroy the armed, logistical, and financial drug trade organizations. Phase I—the first year (2000)—focused military, police, and judicial efforts on Putumayo department and the south. Phase II—the second and third years (2001–2002)—concentrated military, police, judicial, and social efforts in the southeastern and the central parts of Colombia. Phase III—the last 3 years (2003–2005)—expanded the "integrated effort" to the whole country.[47]

During the execution of the three-phased counterdrug plan, protection of human rights was identified as the first priority. Under roles and missions, each counterdrug force had its priorities assigned: COLMIL—insurgents or guerrillas, illegal self-defense or *autodefensas*, drug traffickers, and organized crime; CNP—drug traffickers, organized crime, petty crime; and Department of Administrative Security (DAS)—economic and financial crime, illegal gained wealth by individual or guerrilla groups.

Although counterdrug remained a CNP function, the COLMIL became involved because of the drug ties to the guerrillas and the *autodefensas*. The "basic elements" for the COLMIL–CNP counterdrug efforts were human rights, air interdiction, interdiction of precursor chemicals, COLMIL support of CNP counterdrug operations, destruction of drug processing laboratories and stockpiles, and coca eradication. The counterdrug strategy sought to bring "all elements of the Police and Armed Forces" to bear on the problem "to eliminate large-scale drug production, end large-scale violence and lawlessness by organized groups, promote respect for human rights, and break the link between armed groups and their narcotics industry support."[48] Plan Colombia acknowledged the linkage between the guerrillas and the *autodefensas* with the drug traffickers. This suggested that to combat one it needed to fight the other.

For the United States, Pastrana's plan meant a new, improved, and bigger counterdrug effort—one that provided the CNP with critical military support to increase its coca eradication operations in southern Colombia. The heart of the US "Plan Colombia" centered on a "Push to the South" into coca-growing Putumayo department led by a US-trained and equipped three-battalion Colombian Army counterdrug brigade with its own UH-60 helicopters. Funding of this program shifted the bulk of US counterdrug assistance from the small DIRAN in the CNP to a new counterdrug brigade—an equally small part of the Army—that would be tied to US counterdrug requirements and in support of CNP eradication operations. Between 7 February 2000, when the Clinton administration requested funding for its "Plan Colombia," and 13 July when the President signed the law authorizing $768.5 million in addition to the $163.7 million previously approved for counterdrug programs, the legislative process modified the "Plan Colombia" package. Congress reduced funding requests for ground-based radars, the riverine program, voluntary eradication, alternative crop development, governance, and environmental programs, but tripled funding for human rights. For a third of the cost, 30 UH-60 helicopters were replaced by 60 UH-II helicopters—twice as many with less lift and less capacity. In addition to the human rights vetting requirement by DOS, Congress placed a limit on personnel—500 DOD personnel and 300 civilian contractors—in Colombia at any time in support of "Plan Colombia." After the legislation passed, officials renegotiated the helicopter package, which became 33 UH-1N, 30 UH-II, and 16 UH-60 aircraft.[49]

In Congressional testimony during the period that the "Plan Colombia" legislation developed, a senior DOD official described the US efforts in Colombia as a "threat based, intelligence driven, counterdrug interdiction strategy." DOD programs included detection and monitoring, riverine operations, ground-based radars, and training of human rights vetted personnel such as the counterdrug battalion. He expressed concern that the COLMIL—"heavy on 'tail' and short on 'tooth'"—was "not optimally structured and organized to execute sustained counterdrug operations."[50] That same month, about a third of Wilhelm's annual Congressional testimony concentrated on USSOUTHCOM's counterdrug campaign and on Colombia. Short-term tactical analysis teams and joint planning and assistance teams provided host nation forces in his area training and assessments annually. In Colombia, the counterdrug battalion—three maneuver companies and a headquarters company—had become operational 15 December 1999 at a cost of $3.9 million for training and $3.5 million for equipment. This battalion would be the nucleus of a counterdrug brigade of three battalions that would be part of a 6-year, three-phased "Push into Southern Colombia": (1) 2 years in Putumayo and Caqueta departments, (2) 2 years in Meta and Guaviare departments, and (3) 2 years in Santander and northern departments. Wilhelm noted that the UH-60 would become the standard Colombian helicopter because of its range, survivability, payload, and versatility. He emphasized improvements made in aerial interdiction—the Colombian Air Force received a few upgraded aircraft—and in the riverine program—the Colombian Navy fielded 25 of an anticipated 45 riverine combat units.[51] The USSOUTHCOM's counterdrug and interdiction programs reflected US policy and the American "Plan Colombia."

In March, former ambassador to El Salvador David Passage offered a critique of, and an alternative to, the US counterdrug and human rights policy in Colombia. Calling it "almost irrational to expect that a country fighting for national survival . . . should be able to quickly or easily achieve the truly prodigious transformation necessary to live up to accepted norms of human rights and civil liberties," he noted the Colombian security force improvements in human rights and, based on his experience in El Salvador, questioned the wisdom of denying assistance for alleged human rights violations rather than requiring additional training and prosecutions. Passage noted the major disconnect between Colombian and US goals and the frustrations and misunderstandings caused by a narrow counterdrug focus. "Colombia's overarching national priority is to reestablish sovereignty, regain control over its national territory, end its domestic violence, and resource economic growth for all its people—but the professed US objective is simply to end illicit drug trafficking to American customers." Passage believed that US policy failed to address four Colombian realities: increased control in rural areas and pressure on urban areas by FARC and ELN guerrillas, increased potential for human rights violations as security force personnel "lash out" at civilians, increased reliance of armed and trained paramilitaries for local security, and increased linkages between drugs and the guerrillas and paramilitaries. He observed that over time, the counterdrug effort had evolved from just supporting DIRAN to CNP training to training drug case prosecutors to training and equipping COLMIL units "exclusively devoted to counternarcotics programs." Despite these changes and increased counterdrug funding for 6 straight years in Colombia, the "simple fact" remained that there had been no reduction in illegal drugs. Passage declared that the United States' $250 billion, 20-year counterdrug effort had "no impact at all—absolutely none" on reducing illegal drug supplies. He concluded, "It will not, repeat not, be possible to constrict the production or trafficking in narcotics so long as Colombia's Government cannot enforce Colombia's law over the whole of its national territory."[52]

Passage not only critiqued US counterdrug and human rights policy, he offered suggestions for the Colombian Army—many applicable to the COLMIL—and lessons for US officials based on his El Salvadorian experiences. After praising Pastrana for selecting capable military leaders, Passage acknowledged the long-term challenges of moving beyond the "professional critiques" of the Army that indicated "examples of incompetence and corruption at virtually every level of leadership . . . [that] go all the way down to ignorance and fear among ill-trained, inadequately-equipped, and poorly-led conscripts at the bottom." First, he suggested that the Army needed to develop a strategy for defeating the guerrillas and the paramilitaries. Second, it needed training and doctrine that addressed small unit operations, joint operations with the other military services and the CNP, and "virtually unheard of" night operations. Third, it needed drastic improvement to its "primitive" intelligence system. Fourth, operationally the Army needed a rapid response force, an aerial delivered special force to attack high-value targets (HVT), increased ground and aerial transport assets, and improved aerial medical evacuation—what President Jose Napoleon Duarte in El Salvador equated to another division in terms of enhanced morale and lives saved. Last, but not least, the Army needed to improve its logistics system and the anticipation of maintenance requirements such as services and repair parts. Passage added that the COLMIL should not spend money on things not needed—fighters instead of close air support and transport aircraft for the Air Force and "blue water" instead of riverine and coastal craft for the Navy. Passage offered three lessons from El Salvador to guide the United States in Colombia. First, the United States made it clear that the war was the host nation's war to win or lose, not a US war to win or lose. Second, the United States helped retrain and support the host nation security forces "despite an appalling human rights record," but it did not engage in combat operations and limited its military presence to 55 trainers—"not, note, 'advisors.'" Third, the United States "used all the means at its disposal to compel . . . significant internal reforms."[53] After years of limited governmental support and the leaner years of decertification, many in the Colombian Army might have agreed with Passage's bleak assessment—particularly before 1999—and many would support some of his suggestions that had been identified in the Army transformation process that began in 1999. Other ideas—such as small unit operations—may not have seemed appropriate for fighting the larger, well-trained, and heavily-armed guerrillas. However, the COLMIL transformation process remained largely a self-help program with limited governmental resources and minimal US assistance beyond counterdrug programs. In 2000, the COLMIL understood—even if the United States did not—that the war it had to win or lose was against the guerrillas: FARC and ELN.

COLMIL Reorganization and "Plan Colombia": 2000

The highly publicized fielding of the first counterdrug battalion raised expectations, exposed disagreements, and led to misunderstandings. First, the 18 UH-1N helicopters had not arrived as scheduled—delaying an initial JTF-S foray into Putumayo. Second, human rights vetting remained a bone of contention with the Colombian Army. Third, although American officials and the press expressed high hopes for the battalion, the FARC considered it a threat to its coca cultivation in southern Colombia, and the Colombian Army saw it as a means to gain further access to US assistance. A former COLMIL commander observed, "I don't think the United States understands the problem. . . . It's absurd to think that 18 helicopters and 900 troops will win this war." During the give-and-take in late 1999 over a COLMIL proposal to organize two additional US-funded, US-trained, and US-equipped counterdrug battalions and a brigade staff, disagreements and misunderstandings surfaced. USSOUTHCOM proposed what one observer described as the "Schwarzenegger" battalion—a large, one-of-a-kind, it-exists-no-where-else, high-tech organization based on a US Ranger battalion. Two other options considered were a modified Colombian Army battalion—similar in structure to the US Korean-era battalions—and the smaller Colombian Army counterguerrilla battalion. Larger battalions required more Army personnel and increased the human rights vetting requirements. A dispute over the design and responsibilities of the two battalions led Wilhelm to threaten withdrawal of US support. Tapias reacted in kind by threatening to withdraw his request. In the end, both sides agreed on a military mission—raiding drug laboratories—and to an organization with smaller 780-man battalions. After its fielding in 2001, the counterdrug brigade's primary mission became the provision of ground security for the NAS-supported CNP eradication operations, followed by raiding. What an American military official described as a "clash of egos"—actually more a clash of different military cultures and priorities—ended with a US official stating, "This is their program, their initiative. We are supporting them." United States "cradle to grave" funding of CNP counterdrug programs appealed to a resource-constrained COLMIL while access to US military training met the Colombian Army's needs.[54] However, the Army counterdrug brigade program remained secondary to fighting the guerrillas.

Within its limited resources, the COLMIL continued its long-term adjustments to the growing guerrilla threat. In the first months of 2000, the Colombian Army reorganized its headquarters in Bogotá. It replaced its traditional E1—personnel, E2—intelligence, E3—operations and training, and E4—logistics staff organization with four directorates: operations, personnel, logistics, and training. This permitted the Army to perform its routine administrative, logistical, and training functions and to become a combat command. The director of operations, a major general, oversaw the Colombian Army combat operations. The operations directorate had three branches: operations headed by a colonel, intelligence headed by a brigadier general, and psychological operations headed by a colonel. The training directorate oversaw education, training, doctrine, and lessons learned. The national education-training center remained responsible for the normal schools and a new national training center trained the professionals—formerly re-enlisted conscripts, but expanded to include civilian volunteers and prior service personnel.[55] The creation of the Army Tactical Retraining Center (CERTE) proved a critical training initiative by providing standardized, reoccurring training for all Army combat units. Eventually units would rotate through a cycle of 3 months of operations, leave time, and a month of training by CERTE cadre at a training facility. Training focused on critical tasks and lessons learned—first week: 3 days of human rights, followed by psychological operations and specialty training; second week: soldier skills and marksmanship; third week: squad, platoon, and company training; and fourth week: a field training exercise incorporating the previous 3 weeks of training. The Fifth Division, released from administrative and training responsibilities, evolved into a combat division with the transfer of brigades from other divisions already stationed in its area of responsibility. The Fourth Division, in whose area the *zona de despeje* and JTF-S resided, completed its transition to professionals that spring.[56]

The guerrillas adjusted their operations in response to Army actions. Guerrilla multicolumn attacks declined during the first half of 2000, but attacks on isolated security force outposts continued throughout the year. On 28 March, for example, almost 300 FARC guerrillas supported by the ubiquitous gas cylinder bombs destroyed Vigia del Fuerte in Antioquia department, the second poorest village in Colombia, leaving behind 36 dead, 10 wounded, and 9 missing. Among the dead—some killed by machetes and others burned and decapitated—numbered 21 policemen, the mayor, 7 civilians, and 2 children.[57] Massacres by the paramilitaries of alleged guerrilla supporters increased during the year. In February, over 300 *autodefensas* occupied the town of El Salado in northern Colombia and settled in for a 3-day session of rape, torture, and killing that ended with over 36 dead and nearly 3,000 displaced. Security force—in this case the Colombian Navy marine infantry—failure to respond led to charges of collusion with the paramilitaries.[58] Reportedly, over 200 massacres— guerrilla and *autodefensas*—occurred during the first half of 2000. The FARC and ELN continued their off-and-on negotiations with the government. In March, the FARC declared "Law 002," which required taxes from anyone with over $1 million in assets or they risked kidnapping. It became common for families of kidnapped victims to travel to the *zona* to pay ransoms. In an attempt to broaden its base in the urban areas, the FARC established a clandestine political movement, the "Bolivarian Movement for a New Colombia," in April. During previous local elections, the guerrillas had attempted to disrupt them through threats and killings. In 2000, they shifted to supporting candidates who through intimidation or political belief could be convinced to share government-provided funding with the guerrillas. Given the lack of permanent government security in many municipalities, mayors had to reach an understanding with the guerrillas, leave their posts, or face death. This FARC effort undercut civil-military trust as security force commanders came to view many mayors as FARC supporters. By the end of the year, almost 90 percent of the municipalities had a permanent guerrilla presence or had suffered from an attack.[59]

Combat operations intensified in July. At the beginning of the month, Brigadier General Mario Montoya began Operation JAGUAR—a JTF-S counterdrug operation in 10 municipalities in southern Caqueta and in northern Putumayo—with roughly a 2,500-man Army force consisting of the counterdrug battalion and five counterguerrilla battalions supported by Air Force helicopters. In what was reported as an "unprecedented operation against drug traffickers" and a test of the counterdrug battalion in its "baptism of fire" some 6 months after its activation, the failure to use the 18 US-provided helicopters reportedly because of a lack of funds for fuel became the story. In fact, the long-anticipated counterdrug battalion operations had been delayed first due to a late delivery of the helicopters and then to a disagreement about funding.[60] This caused a problem in May when most of the DOS-funded third-nation contract pilots lost their funding. After President Clinton signed "Plan Colombia" in late July and the United States assumed "cradle to grave" funding for a counterdrug brigade and its helicopters, the DOS agreed in August to fix the problem and in October the DOS estimated that it would take another 3 months to field the contract pilots—almost a full year after the first counterdrug battalion became operational.[61] In mid-July, the FARC launched another series of multicolumn attacks against police stations near the *zona* highlighting the Army need for greater mobility and better intelligence. Once again, the Colombian Army blunted the offensive and forced the FARC back into their base areas after local successes.[62] By mid-2000, the Colombian Army had developed a better understanding of the guerrillas and began to think about controlling the mobility corridors used by the guerrillas to move forces and illegal drugs between their base areas from which attacks were launched throughout the country and between their rear area in southeastern Colombia from which the guerrillas supported and financed their operations.

In late summer, both the Colombians and the Americans made changes in key personnel. Routinely, Colombian security force commanders served 2-year terms. Despite his early relationship with the COLMIL, in August Pastrana retained his COLMIL, Army, and Air Force commanders—Tapias, Mora, and Velasco. He appointed Admiral Mauricio Soto as Navy commander and Major General Luis Ernesto Gilbert as CNP commander. That month, Ambassador Anne W. Patterson arrived in Colombia to replace Kamman. On the military side, in September General Peter Pace—a fellow Marine—replaced Wilhelm who had completed a 3-year tour as USSOUTHCOM commander. Working with the Colombian leadership, Patterson and Pace would manage the beginnings of the US "Plan Colombia."

Colombian legislation reinforced on-going military reform efforts. On 12 August, a new military penal code required human rights cases to be tried by civilian courts and the removal of the military legal system from the chain of command to professional military judges. At the same time, the COLMIL began its efforts to create a Judge Advocate General Corps. On 14 September, a reformed military personnel law permitted the dismissal of officers who had less than 15 years' service—previously forbidden by law without legal confirmation. This allowed the COLMIL to release military personnel credibly accused of gross human rights abuses or with known ties to the *autodefensas*.[63] In October, with unanimous COLMIL approval, Minister Ramirez used this law for the first time to dismiss 388 Armed Forces members—including 89 officers from the military services—to improve the efficiency of the security forces. Despite these legal improvements, manpower and resource constraints continued to limit the COLMIL capabilities. Although a record 48 percent of COLMIL personnel participated in combat operations, 13 percent guarded infrastructure, 6 percent secured the borders, and 33 percent remained in training or in military installations.[64]

Rejecting charges of an abusive military, Tapias acknowledged that some officers might have failed to respond to calls for assistance and that ex-soldiers had joined the *autodefensas* for the pay because of record unemployment in Colombia. However, Tapias maintained the fundamental problem remained "we don't have the capacity to act." For a mountainous and jungle country over six times the size of South Vietnam, the Colombian Army had 17 helicopters and about 55,000 combat soldiers to confront half as many guerrillas and *autodefensas* dispersed throughout the countryside—a challenge for the best soldiers in the world even with perfect intelligence and access to additional helicopters. Restrictions on United States-supplied counterdrug assets in Colombia complicated the problem and increased frustrations. For example, in July a 14-man CNP post held off 300 guerrillas for 27 hours with 3 counterdrug UH-60 helicopters supporting the DIRAN only 20 minutes away. After the policemen ran out of ammunition, they surrendered and were executed. The situation, and others like it, had not been considered by the US approving authorities as an emergency mission.[65]

In the fall, the US Government released its "Plan Colombia." The CNP remained the lead agency under the DOS with a Colombian Army counterdrug brigade providing support. Increased US support focused on the "common counterdrug objective" and would "not support Colombia counterinsurgency efforts." Training, equipping, and supporting a counterdrug brigade—composed of the operational counterdrug battalion, a second counterdrug battalion that began training in August and would become operational in December, a 12-man brigade staff that would become operational in early 2001, and a third counterdrug battalion being formed from vetted Army personnel that was scheduled to begin training in January and become operational in April—with a helicopter package of 13 to 16 UH-60, 33 UH-1N (18 previously delivered), and up to 30 UH-II formed the core of the US counterdrug package. The plan provided some funding for CNP operations; counterdrug intelligence enhancements; interdiction efforts; and judicial reform, human rights, alternative development, and governance programs.[66]

Given the increased military involvement in "Plan Colombia," a civilian DOD official provided Congress an overview of its plan. Noting over 10 years of DOD support for counterdrug efforts in Colombia, he declared the program would build on past programs that had proven successful in Peru and Bolivia. The program had two parts: (1) support for the push into southern Colombia, and (2) support for interdiction efforts. To support the first part, the key tasks became training and equipping the additional counterdrug battalions; establishing an operational brigade staff in February 2001 to conduct counterdrug operations in "the world's largest coca cultivation center" in the departments of Caqueta and Putumayo; building an army aviation support infrastructure for the mix and number of DOS-supplied helicopters; enhancing the counterdrug intelligence collection capability beginning in April 2001; and managing a contractor-led effort to support organizational military reform at the MOD and military service-level that addressed military planning, air and ground logistics support, counterdrug military doctrine, counterdrug military strategy, manning the military, intelligence integration, and command, control, and computers. For the second part, DOD support included modifying two C-26 aircraft with air-to-air radars, forward looking infrared radars (FLIRs) for night operations, and communications equipment by the summer 2001; installing FLIRs in two AC-47 aircraft by the summer 2001; establishing a ground-based radar at Tres Esquinas by October 2001; providing a modern, operationally effective nationwide radar command and control system in Bogotá by fall 2001; continuing the multiple country Andean Ridge intelligence collection program; and initial planning for a ground interdiction—road control—program. Acknowledging that there was "nothing new here for DOD" and that it would "not be easy" to attack the cocaine center in southern Colombia, the official offered three DOD concerns. First, the COLMIL—limited by resources, training procedures, and lack of joint planning and operations—needed to be restructured for success against the drug threat for which DOD had provided a "small portion" of the funding. Second, human rights improvements—from US legislative-mandated requirements, personal example of US trainers, and the reforms of Pastrana—had reduced allegations against the military by 95 percent in 5 years to less than 2 percent of the violations in 1999, but violations by the guerrillas and the paramilitaries provided a "call to action" requiring resources and training for Colombia to which "Plan Colombia" provided a US contribution. Finally, addressing the counterdrug and counterinsurgency issue, the targets of this program, individuals and organizations engaged in narcotrafficking and those armed elements that resisted counterdrug operations, would be engaged—whether "narcotraffickers, insurgent organizations, or illegal self-defense forces." He ended by stressing "numerous restrictions, constraints, and reviews" from the Embassy and USSOUTHCOM to the Joint Staff and DOD minimized the risk to US military personnel and included on "each and every deployment order . . . in no uncertain terms, that DOD personnel are not to accompany host nation personnel on operational missions."[67]

Two General Accounting Office (GAO) reports released that October addressed the implementation challenges—for both the US and Colombia—presented by "Plan Colombia" and one concluded that it would "take years to produce results." On the planning side, no one—DOS, DOD, US Agency for International Development (USAID), or Colombia—had developed detailed implementation plans. However, initial estimates indicated a need for increased Embassy staff—24 in NAS and 40 for USAID—for whom modular offices on the Embassy grounds would be unavailable for at least a year. Economic and social programs would build on pilot projects not yet developed—much less implemented—and requiring years to evaluate. In addition, initial planning had been hasty and uncoordinated—between agencies and between governments. Take DOD for example. For years, the MILGP had assisted specific counterdrug programs such as the riverine and aerial interdiction programs, had little reason to expect major changes, and had focused on the day-to-day management of its on-going counterdrug programs and human rights vetting. The COLMIL had no counterdrug military plan. The USSOUTHCOM lacked a detailed assessment of COLMIL requirements. This caused USSOUTHCOM to base its initial input of COLMIL requirements on incomplete information and on "intuitive assessments of the Colombian Military's basic needs." To remedy this shortcoming, DOD had initiated two studies—one for an aviation support package and another to look into modernizing and restructuring the COLMIL—to identify "operational doctrine, structure, and systems" to make the counterdrug assistance effective.[68] Sharing these planning and coordinating shortcomings with DOD, Colombian and US agencies continued to develop the details of their programs.

Even if the planning had been perfect, there remained significant management and coordination challenges among US agencies, within the Colombian Government, and between the two governments that threatened the prompt execution of programs—thus producing confusion, misunderstanding, frustration, and mistrust. First, the United States had "not always provided the necessary support to operate and maintain the US-provided equipment to the extent possible [needed] to help." Longstanding problems—many encountered in the 1998 assistance package—included inadequate DOS understanding and funding for helicopter support, a lack of helicopter spare parts, a shortage of trained Colombian helicopter pilots and mechanics, DOD inability to provide all the equipment requested, DOS inability to get CNP to assume responsibility for aerial spray eradication, and lack of program oversight—in May 2000, the CNP had its first audit in 15 years.[69] Second, concerns about the ability of Colombia to finance and execute Plan Colombia arose. Not only had international funding fallen below expectations, Colombian attempts to float $1 billion in bonds had raised only $325 million by August with another $325 hoped for by December 2001. Its detailed action plan for Plan Colombia, finally issued in September, had required assistance from a special US interagency task force created in July. Third, governmental policies, bureaucratic procedures, and delayed decisions coupled with long lead times all added to the friction. Training could be accomplished easier than equipment could be acquired. As an example, take the helicopter package for the counterdrug brigade scheduled to be operational in April 2001. In September 2000, DOS estimated it would take over 2 years to complete the delivery of the UH-II helicopters, and that they would be delivered in increments beginning in mid-2001. If an agreement on the specifications for the UH-60 helicopters permitted a contract by December, DOS estimated initial delivery in mid-2001 with final delivery in December. If the contract was not signed then, the final delivery estimate became October 2002—3 years into the 6-year Plan Colombia. Human rights vetting remained another thorny issue that threatened Plan Colombia as the United States maintained the "strictest human rights standards." Training for the second counterdrug battalion had been delayed by the Embassy until the Colombian Army removed a captain accused of a human rights violation even after the Colombian Government had investigated and cleared him of any wrongdoing. Some in the Colombian Army viewed the DOS insistence in removing this officer as an act of personal injustice and a professional—if not national—affront.[70] Expectations for quick and dramatic counterdrug results faded as timelines lengthened and disagreements increased.

Although the FARC had attempted "bait and ambush" operations in 1999 and 2000, the Fourth Division had recognized the gambit and turned the tables on the guerrillas with good results. However, in mid-October in a series of FARC and ELN coordinated attacks, a 500-man FARC force attacked the town of Dabeida in Antioquia department north of Medellín astride a mobility corridor used for arms and drug smuggling. After driving the police from the town, scattering the population, and opening up the corridor, the guerrillas held the town as bait and ambushed a 4th Brigade reaction force. The resulting 3-day battle with 5 FARC fronts left 54 soldiers and 2 policemen dead and a crashed Colombian Air Force UH-60, which had tried to land in the daylight.[71] The guerrillas retained the capability—under the right conditions—to damage Colombian Army units. However, under normal conditions the Colombian Army had proven capable of handling larger guerrilla units. In mid-November, the 5th Brigade in northeastern Colombia intercepted the elite FARC 360-man Arturo Ruiz column—making a 700-mile, 10-department journey by foot, canoe, and truck from southern Colombia to the central Magdalena River valley—on the barren, frigid Berlin plateau about 12,000 feet up in the mountains. In a month-long series of engagements known as Operation BERLIN, Brigadier General Martin Carreno's 1,500-man Colombian Army force suffered 1 killed as it scattered the FARC column—killing 46 and capturing 77—in what was called "the army's shining moment for the year." During this operation, the Army documented FARC use of child soldiers—32 of the prisoners were from 14 to 17 years old and 20 of the killed were children. When the operation finally ended in early 2001, the FARC had lost 71 killed and 132 captured.[72]

After 2 years of negotiations, peace appeared no closer than at the beginning. On the positive side, Plan Colombia provided some international support—particularly in the counternarcotics arena—and an improving COLMIL continued to oppose guerrilla violence. However, during that same 2-year period, conservative estimates indicated that the guerrillas had increased in record numbers: the armed FARC from 11,300 to 16,500, the ELN from 3,500 to 4,500, and the *autodefensas* from 4,500 to 8,100.[73] Violence and lawlessness reigned in large parts of the country displacing 2 million internally and causing over 800,000 to leave the country. An American analyst compared the contrast between urban and rural Colombia to "a sophisticated South American Milan attached to a brutal South American Congo." Despite years of experience and increasing US involvement, Americans had retained "an appalling ignorance of Colombia's hellishly complicated realities" and of its "endemic violence [that] bewilders and dismays many Colombians." Noting that most Colombians viewed the drug war as "America's war" in which they were victims, the lack of financial support for Pastrana's counterdrug-focused Plan Colombia made some sense. The principal complaint of COLMIL officers was the lack of national support to fight the guerrillas. Other criticisms spoke to "outmoded US counterinsurgency doctrine, unreasonable human rights constraints, and weak-kneed politicians." Addressing concerns about the United States being dragged into a quagmire, the analyst observed, "Unless we were to replace the country's nationalistic officers with English-speaking toadies, dependent on US assistance and approval, the Colombian Armed Forces will continue to fight their own wars and resist US military interference. This will be no Vietnam." He concluded that the increased US counterdrug assistance would do little to reduce illegal drugs and would not reduce the amount of violence in Colombia.[74]

Strengthening COLMIL: Plan 2001 and the *Brigada Contra el Narcotrafico* (BRCNA)

At the beginning of the year, COLMIL began a process that led to the development of Plan 2001—the Colombian Army's first campaign plan. On 17 January, the COLMIL held a conference to review the security status, evaluate counterguerrilla and antiparamilitary plans, and consider the "Plan Colombia" guidelines for Army counternarcotics units. Before the meeting, the five new division commanders each met with their brigade commanders to analyze the public order situation in their areas of responsibility, determine regions at risk, and decide on missions for the year. Senior COLMIL commanders then met with the new Army director of operations, Major General Carlos A. Ospina, and the division commanders to discuss topics that included Operation BERLIN, paramilitary actions, car bomb attacks, infrastructure security, and the recently activated *Brigada Contra el Narcotrafico* (BRCNA)—the US-supported counterdrug brigade. The Armed Forces commander briefed the division commanders on the antiparamilitary committee.[75] At a second session later that month, the COLMIL decided to concentrate in 2001 on improving combat capabilities and accomplishing strategic goals; improving security in five departments—Cundinamarca, Antioquia, and Valle where the three largest cities (Bogotá, Medellín, and Cali) were located, Putumayo where JTF-S conducted counterdrug missions, and Arauca where a critical oil pipeline existed; continuing reorganization and modernization of the Armed Forces; meeting the division and brigade objectives in their areas of operation; and creating new elite units. To maintain the proficiency of combat units, the COLMIL leaders committed themselves to maintaining the retraining program between combat operations. To expand the Colombian Army offensive capability, four divisions—First through Fourth—would each activate a BRIM and the Fifth Division would activate a new high mountain battalion from over 10,000 professionals beginning that spring.

An official summarized, "We spent last year reorganizing and modernizing the troops. This year we shall strengthen them."[76]

By 2001, the Army had developed a strategy to counter the guerrillas. First, the Army secured key areas, resources, and infrastructure critical to maintaining the nation. Capitalizing on the lack of popular support for the guerrillas, the Army then focused on the armed guerrilla units and the areas that supported them. In 2000, the Army began concentrating on mobility corridors—areas with limited to no government presence through which the guerrillas moved units, illegal drugs, and weapons between guerrilla base areas or, in Army terms, "generators of direct action," and between guerrilla rear or support areas—the Army's "generators of actions." In 2001, the divisions assumed responsibility for contesting the mobility corridors. The three-BRIM FUDRA and the now separate SF Brigade, the Colombian Army's reaction forces, began to pressure base areas from which the guerrillas launched attacks and JTF-S, with the newly operational BRCNA, attacked FARC support in Putumayo and Caqueta.[77] To implement the strategy, the Army published its first strategic plan, Plan 2001. Within the strategic design, other plans or programs developed the necessary tools. Plan 10,000, a 3-year effort, replaced 10,000 regulars with high school diplomas (noncombat soldiers) each year with 10,000 professionals. By 31 December, the Colombian Army would have over 53,000 professionals—44 percent of its strength. These professionals supplied personnel for the high mountain battalion, additional counterguerrilla battalions, and new urban counterterrorism units (AFEUR). *Plan Fortaleza* or Strength Plan provided 10,000 regulars to replace 10,000 professionals securing critical installations identified in the infrastructure security plan, *Plan Especial Energetico Vial* (PEEV). These professionals provided the personnel for the four divisional BRIMs.[78] In the absence of a national strategy beyond negotiating peace and the vision offered in Plan Colombia, the 2001 military strategy became a guiding concept for a resource-constrained military in its efforts against the guerrillas.[79] No matter how much the COLMIL improved, it lacked the mobility required to react promptly to attacks and it remained too small to secure areas cleared of the "generators of violence."

On paper, the 2,300-man BRCNA, the US-supported counterdrug brigade, with its 77-helicopter support package appeared to be a great asset for the Army. Its three large counterdrug battalions had roughly the strength of the FUDRA, the Colombian Army's three BRIM strategic reserve, and its helicopter package equaled the combined assets of both the Colombian Army and Air Force.[80] In reality, things were not that straightforward. First, US human rights vetting requirements of regular and professional soldiers caused delays and frustrations for both sides. Second, when a two-battalion BRCNA and its 12-man joint staff—comprised of members from each military service and the CNP—became operational in early 2001, only the least capable helicopters had arrived—33 UH-1N. Given the conditions in Colombia, each UH-1N could carry about half a dozen soldiers compared to over twenty for the UH-60. By the time the third counterdrug battalion completed its 18-week training program in late May, the 3-battalion BRCNA had received no additional lift assets—the 14 UH-60 package arrived between July and December and the 30 UH-II package arrived between December and May 2002. The helicopter package completed delivery a full year after the 3-battalion BRCNA became operational.[81] Third, BRCNA counterdrug operations were tied to supporting CNP coca eradication operations. Fourth, counterdrug restrictions combined with the human rights vetting requirements for Army units working with the BRCNA in effect created a long-term, US-interagency-involved "military within the military, a special 'drug fighting' component." Assigned the BRCNA, several Army brigades, and some riverine units in the area, JTF-S (still commanded by Montoya) assumed control of a specially created counternarcotics area in Putumayo and Caqueta departments— an area about the size of Pennsylvania. There, JTF-S supported the US-sponsored, CNP-led "push into southern Colombia" that concentrated on coca eradication. (See Figure 8, Joint Task Force–South (JTF-S) area of operations.) This arrangement permitted the Colombian security forces in the remaining 30 departments to address Colombian security issues with minimum US involvement.[82]

In 2000, as one of two studies to better define COLMIL requirements, the DOD contracted a 14-member team headed by a retired US Army general officer to provide advice to the MOD for COLMIL and CNP reforms in planning, operations, intelligence, training, logistics, and personnel areas. DOD hired the team "not because it has any special expertise," but because USSOUTHCOM "cannot spare 14 men to send to Colombia." For whom the contractors worked—the United States or Colombia—became an issue when their task to recommend "legislation, statues, and decrees" became public knowledge. The DOS indicated that the United States was not trying to "ram military reform down the throats of the Colombians" who could accept or reject any recommendations. In fact, Colombian security sector reform had been ongoing since early 1999.[83] In response to Colombian pressure, in February 2001 DOD decided not to renew its contract. Why? Hired by DOD to come up with ways to "fix a defective soldier" and a broken MOD, the contract team spoke no Spanish, had little to no Latin American experience, and had subject matter experts

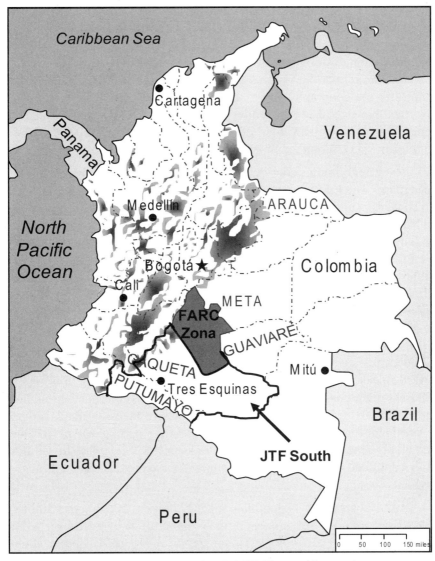

Figure 8. Joint Task Force–South (JTF-S) area of operations.

mostly with Pentagon and European experience. Colombian officials found the final products of little use and not suited to their needs. Manuals contained generic, commonplace observations based on past US military operations rather than specific advice tailored to the current Colombian situation. In the midst of a war, combat-experienced COLMIL general officers had little time for retired US officers who did not understand or failed to address their problems. In fact, the biggest Colombian frustration was that the contractors focused on the drug war and not their war against

73

the guerrillas. In an effort to downplay the situation, the MOD noted, "In a country at war, there is not a lot of time to go to committees" and that the contractor-provided products would serve as "an element of analysis and discussion." An expert on Colombia observed that some US officials tend to "lord it over the Colombians, they don't know the language, or the culture or the geography, and they certainly don't know the interservice rivalries that have been impediments to reform." DOD had no plans to replace the contract team as MOD officials indicated they would continue to reform "a la Colombiana"—Colombian-style.[84]

Peace negotiations continued in what one called a "dialogue without a road map." The European trip in 2000 with FARC representatives brought no consensus and increased FARC political prestige in parts of the world. In November, the FARC walked out of the talks accusing the government of failing to combat the *autodefensas*. During the same time, talks with the ELN about a *zona* continued. In February 2001, residents in Santander protested a proposed government *zona* for the ELN as the AUC attacked ELN areas. On 8 February, Pastrana met with Marulanda in the *zona* and agreed to extend it for another 7 months, to denounce the paramilitaries as a common enemy of the FARC and of the government, and to permit foreign diplomats to observe the next talks. On 8 March, observers from 22 countries attended the talks in the *zona*.[85] Some Colombians saw the FARC release of 242 police and military captives in June as a sign of progress in the peace process.[86] However, as the talks continued throughout the year, both the FARC and the ELN conducted attacks on small, isolated government forces and on critical infrastructure—particularly power lines and oil pipelines, but no major offensives as in 1999 and 2000. In addition, both continued their kidnappings, summary executions, and killings of civilians.[87]

In the absence of Colombian security forces, *autodefensa* military strength continued to grow. Armed paramilitary presence in the municipalities reached over 40 percent by the end of 2001. Human Rights Watch referred to the paramilitaries as the "Sixth Division" in the five-division Colombian Army.[88] The Armed Forces commander predicted in February that the paramilitaries would outnumber the guerrillas in 3 years because "they have more support from civil society."[89] Despite its illegal status, the AUC had a 15-percent favorable rating—five times greater than the FARC. In an interview, Carlos Castano—the AUC leader— noted the *autodefensas* did not "represent the best solution to Colombia's problems." But they were "one, perhaps the only one, and one that the Colombian people see at this moment." He added the Armed Forces "are like brothers. Our enemy is the guerrillas and that has not changed." In defense of AUC killings and atrocities, Castano said, "This is an irregular conflict. You have to understand that the guerrillas, not us, determined the conflict's characteristics."[90] From January through April, the AUC conducted a successful offensive to displace the ELN in northeastern Santander department where, in the words of an observer, "The AUC, and not the Colombian Armed Forces, is the main reason the ELN is on the ropes militarily." During this fighting, over 180 civilians died and about 4,000 departed the area.[91] A farmer simply explained, "The army comes, the army goes. The [armed] groups remain to tell us what to do."[92] This absence of government—a defining attribute of the war—created the void in which armed irregular groups operated.

Despite increases in the number of professionals, in training improvements, and in combat operations, the Colombian Army end strength remained too small to counter the growing guerrilla threat, much less defeat it. Fifty times the size of El Salvador, Colombia had an Army roughly twice the size of that in El Salvador during its civil war. Minister of Defense Gustavo Bell, the vice president and government human rights advocate who replaced Ramirez in May, noted that military operations had retaken some critical mobility corridors. However, commanders acknowledged their inability to establish a permanent government presence meant that the guerrillas returned after they withdrew from an area. This increased the likelihood that the security forces would, in the words of a villager, "see the whole population as guerrillas." Because negotiations remained the government priority, the MOD received about 7.5 percent of the budget—below its 1990 level—but funding rose from $1.5 billion in 1995 to over $1.9 billion in 2001.[93] Generating additional combat power through helicopter mobility proved difficult because of inadequate numbers, counterdrug restrictions, and five separate helicopter fleets listed in decreasing size—CNP, JTF-S, Air Force, Army, and Navy.[94] To strengthen the military forces, in August President Pastrana signed the MOD-recommended National Defense and Security Law—the first substantial reform to the national security law since 1965. In areas declared by the President to be "theaters of operations," the military could establish martial law and replace civilian rule. The military received the authority to arrest and investigate if the *fiscalia*—criminal investigators— were not available. The automatic release of prisoners not delivered to the *fiscalia* after 36 hours no longer applied, but delays in processing required justification. The law, modified by the legislative process, reduced the time allowed for the military to investigate human rights violations from 12 to 2 months.[95] Without major political changes, the military forces continued to be bound by their legal and resource-constrained straightjacket.

At the beginning of the George W. Bush administration, US policy toward Colombia remained basically unchanged. Counterdrug and human rights drove the relationship. The administration changed the limits on DOD Plan Colombia personnel from 500 DOD personnel and 300 US contractors to 400 each—a total of 800.[96] In September, the DOS had designated the AUC a foreign terrorist organization as United States pressure for action against the *autodefensas* continued. Human Rights Watch reported that all CNP, DIRAN, Air Force, and Navy units had been vetted for US aid compared to only 11 Army units—all related to JTF-S, SF, or the aviation brigade.[97] What many viewed as the Army's inability to meet human rights standards in fact reflected an Army efforts to minimize its interface with the US counterdrug effort. Simply put, why vet units that received no US assistance? During this period, human rights training continued to receive Army emphasis. Many Colombians had come to view the "human rights cartel"—as they called it—as just another guerrilla weapon for attacking COLMIL leaders.[98] By year end, the counterdrug program appeared to get underway. The CNP, supported by JTF-S, began phase 1 of the counterdrug program almost 2 years later than planned. Aerial spraying in Putumayo almost doubled all spraying in Colombia in 2000. JTF-S supported these eradication missions and conducted over 200 operations against drug labs and infrastructure. Other than the Air Bridge Denial Program, which had been suspended after the shoot down of a missionary airplane in Peru in April, counterdrug programs grew. The DIRAN increased its airmobile interdiction units—its elite *Junglas*—from one to three companies, the Colombian Army began talks about a second counterdrug brigade, and Navy riverine forces reorganized with a 5-year goal of 25,000 personnel in two brigades.[99] After years of effort and innumerous delays, the counterdrug program in the south had begun.

Collapse of the Peace Process: 2002

By the beginning of 2002, the years of peace efforts had run their course. Pastrana, unable to force a settlement on the guerrillas, had lost support for negotiations; the guerrillas had missed an opportunity for a settlement but had used the time to improve their capacity to continue the military struggle; and the *zona de despeje* had come close to disestablishment in late 2001. On 10 January, President Pastrana announced the end of peace talks and gave the FARC 48 hours to evacuate the *zona*. He failed to indicate when the time limit began or what he intended to do if they did not evacuate. Immediately, the COLMIL increased its readiness and began to move forces toward the *zona*, in the words of a military official, "to be ready for whatever decision the President makes, whether he says go back home or go in."[100] The COLMIL concentrated almost 13,000 personnel, to include non-BRCNA units from JTF-S.[101] By this time, full-time guerrilla strength had grown to almost 22,000 with a presence in over 90 percent of the 1,097 municipalities. The two largest groups—the FARC and the ELN—fielded over 17,000 and over 4,000 respectively. Along with the guerrillas in 2001, the AUC had grown almost 25 percent to 10,500.[102] Despite these increases, the COLMIL end strength had remained basically static, even with the reforms and an Army reorganization that incorporated the units raised in 2001—four divisional BRIMs, additional divisional counterguerrilla battalions, a mountain battalion, eight infrastructure security or PEEV battalions, and the BRCNA.[103] As Pastrana considered his next action, FARC and ELN attacks on civilians and infrastructure increased throughout the country. To meet these multiple threats, the COLMIL commander called his division and brigade commanders together in early February for the second time in weeks to review operations to counter narcoterrorists—a term increasingly used after 11 September 2001 for guerrillas and *autodefensas*—operations countrywide.[104] The hijacking of an airplane and taking hostage of a prominent senator proved the final straw. That night, in a nationwide address, Pastrana ordered the reoccupation of the *zona*. Referring to the FARC as terrorists for the first time in his presidency, he warned, "Hard times are coming, without a doubt."[105]

At midnight on 20 February, the Colombian Air Force, acting on its technical intelligence confirmed by human intelligence, bombed FARC camps and airstrips to begin Operation THANATOS—what would become more than a 3-month operation in the former *zona de despeje*. The task of reoccupying an area the size of Switzerland with 13,000 military personnel fell to Major General Gabriel E. Contreras. Air Force aircraft and Navy riverine units supported the Army units involved—brigades from the Fourth Division, four counterguerrilla battalions, and the Army reaction forces—the FUDRA and the SF Brigade. Anticipating the attack, the FARC had dispersed into the countryside and offered minimal resistance in the former *zona*. On short notice, the Army executed its plan without major difficulties. Conducting over 300 military operations in the first month to reoccupy the former *zona*—in addition to counterguerrilla operations elsewhere—drained COLMIL funding and exhausted the Air Force's stockpile of 500-pound bombs.[106] Not only were funds stretched, General Tapias warned, "It is important that everyone understand that we do not have troops available or reserves to confront emergencies" and the public "cannot expect . . . total control of areas larger than El Salvador . . . when . . . the FARC has been present in [some parts] . . . for more than 30 years."[107] Years of government failure to anticipate—and thus prepare to combat the guerrillas—would have to be made up by continued COLMIL adaptation.

As guerrilla attacks—now small-unit hit-and-run affairs, car bombings, terrorist incidents, and kidnappings—increased countrywide after the closure of the *zona*, Pastrana took several actions. First, he established a "theater of operations"—the former *zona* and areas to its north just south of Bogotá that included 19 municipalities in six departments—to prevent guerrilla attacks on infrastructure supporting Bogotá similar to attacks that left over 60 urban areas without electricity, water, or phone services or isolated by guerrilla roadblocks or damaged bridges. In this designated region, the military could establish curfews, create roadblocks, register civilians, and set business hours. If disputes arose, the mayors could appeal to Pastrana for a decision.[108] Second, Pastrana publicized FARC narcotrafficking activities. Satellite and aerial photography indicated and the CNP confirmed after reoccupation of the *zona* that coca cultivation there had increased 2.5 times and poppy cultivation had started. In 2002, 12 percent of the illegal narcotics cultivation in Colombia existed inside the former *zona*—the result of FARC drug ties, exemption from counterdrug operations, and CNP eradication efforts in Putumayo, which pushed growers into other parts of the country.[109] Third, Pastrana strengthened the COLMIL by providing $110 million for continued military operations and by increasing the Colombian Army by 10,000 regulars a year for 3 consecutive years.[110] Fourth, with the collapse of the peace process, Pastrana appealed to the United States for increased military assistance beginning with improved intelligence sharing, a request approved almost immediately, and the lifting of the legislative-mandated counterdrug restrictions on the BRCNA. At the time, a former MOD official warned, "Intelligence is only as good as your ability to act on it. What the Americans will do is not the key to success here. There are many other things that need to be done first. It's going to cost a lot of money and it's going to cost a lot of lives." US estimates suggested the need for doubling the Colombian Army strength just to slow the guerrilla growth. Tapias believed he needed a quadrupling of the Army to "restore . . . normalcy"—to secure all the major towns, to control the roads, and to eradicate 50,000 acres of coca that assisted in financing FARC.[111] The collapse of the peace process, the reoccupation of the *zona de despeje*, and the appeal for American assistance occurred at a time when US policy on counterdrugs in Colombia had begun to shift toward counterterrorism.

Before the collapse of the peace process in mid-February, the Bush administration had expressed concern about Colombian infrastructure security—specifically the protection of the Cano Limon-Covenas oil pipeline in the northeast. Then on 21 March, responding to the Colombian appeal for assistance, the administration requested a counterterrorism supplemental bill, which included "new authorities" that permitted the United States "to address the problem of terrorism in Colombia as vigorously as [it] currently address[ed] narcotics" and to assist Colombia to "address the heightened terrorist risk that has resulted" with the closure of the *zona de despeje*. Accepting linkage between the narcotraffickers and terrorists in Colombia, the bill sought to remove the counterdrug restrictions in the fight against narcoterrorists by permitting the units trained with and the equipment provided by the $1.7 billion US counterdrug assistance since July 2000 to be used to attack the narcoterrorists, which were US-designated foreign terrorist organizations (FTO): FARC, ELN, and AUC. The DOS human rights vetting requirements to qualify for US assistance and the DOD 800-personnel cap remained unchanged. The $35 million counterterrorism legislation, signed into law in late summer, included $25 million in Non-proliferation, Anti-terrorism, Demining, and Related Programs (NADR) for CNP and COLMIL equipment and antikidnapping training; $6 million in foreign military financing (FMF) to begin training for a Colombian Army infrastructure security brigade for the Cano Limon-Covenas pipeline—the first substantial noncounterdrug funding for the COLMIL; and $4 million in International Narcotics Control Law Enforcement (INCLE) funding for CNP programs to reestablish a presence in abandoned municipalities. (See Table 3, US assistance to Colombia, 1999–2002, for funding during the Pastrana presidency that includes the supplemental funding for Plan Colombia and fiscal year 2002 in bold italics.) This proposed legislation "to mount an effective campaign against terror" rather than counterdrugs marked a major shift in US policy toward Colombia.[112]

Having replaced General Pace on 1 October 2001 as acting USSOUTHCOM commander, Major General Gary Speer provided the annual update to Congress on 5 March. In Colombia, he noted the closure of the *zona* and highlighted the fact that the Cano Limon-Covenas oil pipeline had not been operational for over 266 days in 2001—almost 9 months—at a cost of $40 million per month. He noted a "steady improvement in the professionalism and respect for human rights" by the COLMIL and increased counterdrug operations. Speer addressed legal assistance projects to develop a Judge Advocate General (JAG) school and to improve human rights and added that "in a short period of time" the COLMIL had "emerged as one of the most respected and trusted organizations" in Colombia with

Table 3. US assistance to Colombia, 1999–2002[113]

Dollars in millions	FY 1999	FY 2000/ Plan Colombia Supplemental	FY 2001	FY 2002/ Supplemental	TOTAL
Economic—USAID	**13**	**4**	**4**		**21**
Developmental Aid	3				3
Economic Support Funds		4	4		8
Disaster Assistance	10				10
Counternarcotics	**287.2**	**954.2**	**259.9**	**518.5**	**2,019.8**
DOS International Narcotics Control	205.9	50 / 768.5*	48	380.5 / 4	1,456.9
DOS Air Wing	30	31.3 / *in above	35	41.8	138.1
DOD Sec 1004: Nonlethal CD/Police	35.9	90.6	150	83.2	359.7
DOD Overlapping Sections 1004/124		6.6	4.6	5	16.2
DOD Sec 1033: Nonlethal Riverine	13.6	7.2	22.3	4	47.1
Administration of Justice	1.8				1.8
Antiterrorism—NADR				/ 25	**25**
Military	**.9**	**.9**	**1**	**7.2**	**10**
IMET	.9	.9	1	1.2	4
FMF Grants				/ 6	6
Drawdowns	**72.6**				**72.6**
DOD Sec 506: Nonlethal Excess	58.1				58.1
Other US Departments Sec 506	14.5				14.5
TOTAL	**373.7**	**190.6 / 768.5**	**264.9**	**515.7 / 35**	**2,148.4**

less than 3 percent of the human rights complaints—compared to 60 percent a few years before—against the security forces. He called the counterdrug brigade "the best trained unit" in the Colombian Army, noted delivery of the "Plan Colombia" UH-60 package, and projected the final delivery of the 30 UH-II package in August—almost 14 months after activation of the 3-battalion counterdrug brigade. The counterdrug brigade success had led to support for a Colombian request to create a second counterdrug brigade in fiscal year 2003. Major engineer projects continued to develop a support infrastructure for riverine, counterdrug brigade, and aviation units. Speer stressed the importance of military education programs such as those offered at the Western Hemisphere Institute for Security Cooperation (WHINSEC) at Fort Benning, Georgia. He concluded with his scarcity of resources—although the USSOUTHCOM FMF program had increased to $8.7 million, it remained less than 0.1 percent of the worldwide program.[114] Colombia remained just one—although an important one—of the many USSOUTHCOM programs that now included detainee operations at Guantanamo Bay, Cuba.

For the remainder of his presidency, Pastrana struggled against an increasingly critical situation. A multiday fight in May at Bellavista in northwest Colombia between 1,400 heavily-armed, well-financed, and unconstrained *autodefensas* and FARC guerrillas to control a critical smuggling corridor caused heavy casualties, "the largest single massacre of civilians recorded," after a FARC propane gas cylinder bomb hit the church killing 117 civilians—about one-third children. Despite warnings, many of which often proved false, this isolated town received no assistance from the thinly stretched security forces.[115] Throughout the country, the demand for security forces continued to exceed the supply. In early May, Pastrana tried another peace initiative, "The Road towards Peace and the Counterterrorism Strategy," that foundered on the heavy fighting between the guerrillas and the *autodefensas* in areas of the country; on the decision of the European Union not to designate the FARC or the ELN as terrorist organizations while designating the AUC as one; and on a bleak DOS assessment of the political and military circumstances. Military officers reacted to a suggestion to create a joint command by calling it a threat to "what is functioning by constitutional mandate."[116] Despite his continued efforts, Pastrana could not accomplish in his last 3 months what he had failed to accomplish in the previous 4 years.

In a 29 June speech to the nation, Pastrana called for national unity in resisting the terrorists who—in his words—had failed to defeat the Army, failed to paralyze the country through attacks on infrastructure, and failed to disrupt the legislative and Presidential elections. Declaring to have strengthened the security forces "like never before in history" by referring to his recent increase in MOD funding and the addition of 10,000 personnel, he noted the FARC had changed to terrorist attacks because of its inability to defeat the Colombian Army. In an effort to attack the FARC leadership, Pastrana announced two programs: first, a rewards program for information leading to the capture of FARC leaders—$2 million for members of the Secretariat and $1 million for bloc commanders; and second, the creation of a "special unit" in the military, in the police, and in the DAS "tasked with finding and capturing the members of the FARC Secretariat." In meeting the current situation, Pastrana and his advisors rejected calls for "a state of internal commotion" as inappropriate, but warned the media not to become "a mouthpiece for the terrorists' threats."[117]

At the 20 July Independence Day military parade, Pastrana reviewed the improvements in security made during his administration—what he called an "unprecedented strengthening of the Armed Forces." To support this claim, he noted the 75 percent increase in fighting soldiers—from 75,000 to almost 140,000; the 150 percent increase in professionals—from 22,000 to over 55,000; the Strength Plan which permitted a 10,000 increase in soldiers each year through 2004; the 400 percent increase in UH-60 gunships—from 4 to 16; the increase in transport helicopters from 75 to 176; and the creation of new units—the FUDRA, four BRIMs, the BRCNA, 24 riverine combat elements, and the JIC. He referred to the improvements in respect for human rights and to the legislative reforms that defined professional careers, health and welfare benefits, promotions, and the military justice system. The CNP improved in similar ways—increased manpower, new programs, and reactivations of rural police stations. Pastrana mentioned pending US legislation that would remove counterdrug restrictions and provide additional funds as a way "to collect resources and tools for our Armed Forces." He concluded by thanking General Tapias, his Ministers of Defense—Ramirez for modernization and strengthening and Bell for human rights improvements, the COLMIL and CNP commanders, and the members of the Armed Forces for their work.[118] In fact, most of the security force improvements during the Pastrana administration had been COLMIL and MOD initiatives. From before the establishment of the *zona de despeje* to Plan Colombia to the collapse of negotiations, Pastrana had focused on the pursuit of peace through talks, not through military strength. As a result, the illegal armed groups—particularly the FARC and the AUC—had increased to record numbers.

Before the end of negotiations in February, peace candidates had dominated the field for the national elections—legislative on 11 March and Presidential on 26 May. Liberal Party leader Horacio Serpa had dominated the 11-person Presidential field for almost a year. With the collapse of the peace process, the Colombian people rejected "peace" candidates. Serpa slipped from leading the polls in January to less than 30 percent support by the end of February. That month, the FARC kidnapped one of the Presidential candidates—Ingrid Betancourt. Another candidate running as an independent, Alvaro Uribe—a hard-line, former governor of Colombia's most populous department, Antioquia—had almost single-handedly called for taking the fight to the guerrillas through firmer action, law and order, a larger military, and arming the population. Attracting little support in 2001, by January 2002 his message had garnered him a 39 percent approval rate. Although the FARC and ELN had attempted to disrupt both elections—to include three assassination attempts on Uribe—over 11 million Colombians turned out on 26 May to give Uribe an unheard of first-round victory with 53 percent of the vote.[119] On becoming President in August, Uribe would face a severe security situation caused in large part by the failure of the Pastrana peace effort. On the other hand, without that failure and the rejection of negotiations by the Colombian people, Uribe would not have been elected.

Notes

1.   Diana J. Schemo, "Colombia Installs New President Who Plans to Talk to Rebels," *New York Times,* 8 August 1998, http://proquest.umi.com/pqdweb?index=4&did=32695228&SrchMode=2&sid=19&Fmt=3&VInst=PROD&VType=PQD&RQT=309&VName=PQD&TS=1236966804&clientId=5904 (accessed 5 November 2008).

2.   "Commander Defends Colombian Armed Forces in Interview," *BBC Monitoring Americas—Political*, 10 November 2006, http://proquest.umi.com/pqdweb?index=0&did=1159878151&SrchMode=2&sid=21&Fmt=3&VInst=PROD&VType=PQD&RQT=309&VName=PQD&TS=1236966958&clientId=5904 (accessed 4 November 2008).

3.   Douglas Farah, "US to Aid Colombian Military: Drug-Dealing Rebels Take Toll on Army," *Washington Post,* 27 December 1998, http://proquest.umi.com/pqdweb?index=5&did=37803016&SrchMode=2&sid=22&Fmt=3&VInst=PROD&VType=PQD&RQT=309&VName=PQD&TS=1236967126&clientId=5904 (accessed 5 November 2008).

4.   Schemo, "Colombia Installs New President Who Plans to Talk to Rebels."

5.   Frank Bajak, "Colombia's New President Purges Military Leadership," *San Antonio Express–News,* 10 August 1998, http://proquest.umi.com/pqdweb?index=7&did=1206820531&SrchMode=2&sid=4&Fmt=3&VInst=PROD&VType=PQD&RQT=309&VName=PQD&TS=1242063078&clientId=5904 (accessed 11 May 2009).

6.   "Colombian Army Says 40 Killed in Clash With Rebels," *Washington Post*, 17 August 1998, http://proquest.umi.com/pqdweb?index=1&did=33165732&SrchMode=2&sid=1&Fmt=3&VInst=PROD&VType=PQD&RQT=309&VName=PQD&TS=1242064952&clientId=5904 (accessed 11 May 2009); Notes from discussion with Colombian officer present at this fight.

7.   Notes from discussions with senior Colombian Military officers.

8.   Paul E. Saskiewicz, "The Revolutionary Armed Forces of Colombia—People's Army (FARC-EP): Marxist-Leninist Insurgency or Criminal Enterprise?" (Monterey: CA: Naval Postgraduate School Thesis, December 2005), 75–78.

9.   Nina M. Serafino, "Colombia: Conditions and US Policy Options," Congressional Research Service Report, updated 12 February 2001 (Washington, DC: Library of Congress), 18; Peter DeShazo, Tanya Primiani, and Philip McLean, *Back from the Brink: Evaluating Colombia, 1999–2007* (Washington, DC: Center for Strategic and International Studies, November 2007), 8; Myles R.R. Frechette, *Colombia and the United States—The Partnership: But What is the Endgame?* (Carlisle Barracks, PA: Strategic Studies Institute, February 2007), 12–13.

10.   "Colombian Soldiers Re-enter Disputed City/Rebels' Attack Raise Doubts About Peace Bid," *Houston Chronicle*, 5 November 1998, http://proquest.umi.com/pqdweb?index=8&did=35718239&SrchMode=2&sid=1&Fmt=3&VInst=PROD&VType=PQD&RQT=309&VName=PQD&TS=1242141204&clientId=5904 (accessed 5 May 2009); Jared Kotler, "At Least 70 Dead

in Colombia Rebel Attack: Assault on the Police Station was Waged Even as Peace Talks Approach," *Fresno Bee,* 3 November 1998, http://proquest.umi.com/ pqdweb?index=2&did=35684454&SrchMode=2&sid=3&Fmt=3&VInst=PRO D&VType=PQD&RQT=309&VName=PQD&TS=1242147398&clientId=5904 (accessed 11 May 2009); David Spencer, "Latin America, FARC's Innovative Artillery," *Jane's Intelligence Review,* 1 December 1999, http://search.janes. com/Search/documentView.do?docId=/content1/janesdata/mags/jir/history/jir99/ jir00594.htm@current&pageSelected=allJanes&keyword=david%20spencer&b ackPath=http://search.janes.com/Search&Prod_Name=JIR& (accessed 4 March 2009); Notes from discussions with senior Colombian Military officials; Brigadier General Freddy Padilla commanded the 7th Brigade at Mitú.

11. US Department of State, "International Narcotics Control Strategy Report for 1998," Colombia Section (Washington, DC: US DOS, February 1999), http://www.state.gov/www/global/narcotics_law/1998_narc_report/samer98. html (accessed 1 October 2008).

12. "Colombia: Commander Accepts Responsibility for Errors, Army's Reorganization," *BBC Monitoring Americas—Political,* 15 November 1998, http://proquest.umi.com/pqdweb?index=0&did=35964886&SrchMode=2&sid= 2&Fmt=3&VInst=PROD&VType=PQD&RQT=309&VName=PQD&TS=1242 143016&clientId=5904 (accessed 5 November 2008).

13. "Colombia: Commander Accepts Responsibility for Errors, Army's Reorganization."

14. US Department of State, "Colombia Country Report on Human Rights Practices for 1998," 26 February 1999, http://www.state.gov/www/global/human_ rights/1998_hrp_report/colombia.html (accessed 22 October 2008).

15. Frank Bajak, "US Trains Colombian Military to Resist Rebels," *Austin American Statesman,* 6 December 1998, http://proquest.umi.com/pqdweb?inde x=0&did=36560244&SrchMode=2&sid=1&Fmt=3&VInst=PROD&VType=P QD&RQT=309&VName=PQD&TS=1242153130&clientId=5904 (accessed 5 November 2008).

16. Farah, "US to Aid Colombian Military; Drug-Dealing Rebels Take Toll on Army."

17. Quoted in Douglas Farah, "Colombian Army Fighting Legacy of Abuses," *Washington Post,* 18 February 1999, http://proquest.umi.com/pqdweb ?index=2&did=39087527&SrchMode=2&sid=1&Fmt=3&VInst=PROD&VTyp e=PQD&RQT=309&VName=PQD&TS=1242237748&clientId=590 (accessed 5 November 2008).

18. Alfred Rangel Suarez, *Fuerzas Militares para la guerra: La agenda pendiente de la reforma militar* (Bogotá: Ensayos de Seguridad Y Democracia, November 2003), 43–69; Notes from discussions with senior Colombian Military officials.

19. Patricia Bibes, "Colombia: The Military and the Narco-Conflict," *Low Intensity Conflict & Law Enforcement* (Spring 2000): 39–44.

20. Nina M. Serafino, "Colombia: US Assistance and Current Legislation," Congressional Research Service Report, updated 15 May 2001, 19.

21. Charles E. Wilhelm, "Statement before the Senate Armed Services Committee," 13 March 1999, http://www.armed-services.senate.gov/statemnt/1999/990413cw.pdf (accessed 1 October 2008).

22. Dean A. Cook, "U.S. Southern Command: General Charles E. Wilhelm and the Shaping of U.S. Military Engagement in Colombia, 1997–2000," in *America's Viceroys: The Military and U.S. Foreign Policy*, ed. Derek S. Reveron (New York, NY: Palgrave Macmillan, 2004), 148–149.

23. Notes from discussion with American civilian and military officials; Jimmy Burns and Adam Thomson, "Shades of Vietnam in Anti-drugs Fights: US Army Officers are Joining Colombia's Action Against Narcotics—But It Will Involve Them in Conflict with Rebel Groups Too," *Financial Times*, 26 October 1999, http://lumen.cgsccarl.com/login?url=http://proquest.umi/com/pqdweb?did= 45794559&sid=2&Fmt=3&clientld=5094&RQT=309&VName=PQD (accessed 5 November 2009).

24. Thomas A. Marks, *Colombian Army Adaptation to FARC Insurgency* (Carlisle Barracks, PA: Strategic Studies Institute, January 2002) 24; Serge F. Kovaleski, "Colombian Army Drill: Respect for Rights: Training Course Tests Empathy, Discipline," *Washington Post*, 29 August 1999, http://proquest.umi.com/pqdweb?index=9&did=44275664&SrchMode=2&sid=1&Fmt=3&VInst=PROD&VType=PQD&RQT=309&VName=PQD&TS=1242403982&clientId=5904 (accessed 5 November 2008).

25. "Colombia's Defense Minister Quits Over Concession to Rebels," *New York Times*, 27 May 1999, http://proquest.umi.com/pqdweb?index=13&did=41912447&SrchMode=2&sid=3&Fmt=3&VInst=PROD&VType=PQD&RQT=309&VName=PQD&TS=1242327099&clientId=5904 (accessed 14 May 2009); "The Warning from Colombia's Generals," *Washington Times*, 30 May 1999, http://proquest.umi.com/pqdweb?index=9&did=41983631&SrchMode=2&sid=3&Fmt=3&VInst=PROD&VType=PQD&RQT=309&VName=PQD&TS=1242327099&clientId=5904 (accessed 14 May 2009).

26. Cynthia Watson, "Civil Military Relations in Colombia: Solving or Delaying Problems?" *Journal of Political and Military Sociology* (Summer 2005): 102.

27. Quoted in Larry Rohter, "As Colombia Declares an Alert, Rebel Offensive Rages On," *New York Times*, 12 July 1999, http://proquest.umi.com/pqdweb?index=3&did=43068777&SrchMode=2&sid=1&Fmt=3&VInst=PROD&VType=PQD&RQT=309&VName=PQD&TS=1242335255&clientId=5904 (accessed 14 May 2009); Marks, *Colombian Army Adaptation to FARC Insurgency*, 11.

28. US Department of State, "Colombia: Country Reports on Human Rights Practices—1999" (Washington, DC: US DOS, 23 February 2000), http://www.state.gov/g/drl/rls/hrrpt/1999/380.htm (accessed 22 October 2008).

29. Human Rights Watch, *The Ties That Bind: Colombia and Military-Paramilitary Links* (New York, NY: Human Rights Watch, 1 February 2000), http://www.hrw.org/en/reports/2000/02/01/ties-bind?print (accessed 23 January 2009).

30.	Kovaleski, "Colombian Army Drill: Respect for Rights."

31.	US Department of State, "Colombia: Country Reports on Human Rights Practices—1999."

32.	Juanita Darling and Ruth Morris, "Crash Points to Military Role of US in Colombia," *Los Angeles Times*, 28 July 1999, http://proquest.umi.com/pqdweb?index=17&did=43516995&SrchMode=2&sid=4&Fmt=3&VInst=PRO D&VType=PQD&RQT=309&VName=PQD&TS=1242413974&clientId=5904 (accessed 5 November 2008).

33.	"Colombia: River Brigade Forces Begin Operations on 4th August," *BBC Monitoring Americas—Political*, 1 August 1999, http://proquest.umi.com/pqdweb?did=43586599&sid=1&Fmt=3&clientId=5094&RQT=309&VName=P QD (accessed 5 November 2008); Ricardo A. Flores, "Improving the US Navy Riverine Capability: Lessons from the Colombian Experience" (Monterey, CA: Naval Postgraduate School Thesis, December 2007).

34.	Larry Rohter, "With US Training, Colombia Melds War on Rebels and Drugs," *New York Times*, 29 July 1999, http://proquest.umi.com/pqdweb?i ndex=6&did=43483369&SrchMode=2&sid=4&Fmt=3&VInst=PROD&VType =PQD&RQT=309&VName=PQD&TS=1242413974&clientId=5904 (accessed 5 November 2008).

35.	Sibylla Brodzinsky, "Colombia Turns down Dilapidated US Trucks: Repairs Too Costly Despite 'Donation,'" *Washington Times*, 2 December 1999, http://proquest.umi.com/pqdweb?index=1&did=46807467&SrchMode=2&sid= 1&Fmt=3&VInst=PROD&VType=PQD&RQT=309&VName=PQD&TS=1242 421733&clientId=5904 (accessed 5 November 1999).

36.	Notes from discussion with senior Colombian Military officer.

37.	"Colombia: President Pastrana's Handling of Military Crisis Assessed," *BBC Monitoring Americas—Political*, 14 December 1999, http://proquest.umi.com/pqdweb?index=3&did=47218556&SrchMode=2&sid=1&F mt=3&VInst=PROD&VType=PQD&RQT=309&VName=PQD&TS=1242419 601&clientId=5904 (accessed 5 November 2008).

38.	DeShazo, Primiani, and McLean, *Back from the Brink*, 9.

39.	Serafino, "Colombia: Conditions and US Policy Options," 21–22.

40.	Marks, *Colombian Army Adaptation to FARC Insurgency*, 11.

41.	Colombian Army, "Disposicion Numero 000010 Por medio de la cual se reorganiza el Ejercito Nacional," 23 November 1999. Document signed by Colombian Army Commanding General Jorge Enrique Mora. This document reorganized the Army in accordance with Article 23 of Law 1932 of 1999.

42.	Marks, *Colombian Army Adaptation to FARC Insurgency*, 13.

43.	Cook, "General Charles E. Wilhelm and the Shaping of U.S. Military Engagement in Colombia," in Reveron, *America's Viceroys,* 149–152; Darren D. Sprunk, "Transformation in the Developing World: An Analysis of Colombia's Security Transformation" (Monterey, CA: Naval Postgraduate School Thesis, September 2004), 41; Notes from discussions with Colombian and American civilian and military officials.

44.	Serafino, "Colombia: Conditions and US Policy Options," 22–23.

45. Republic of Colombia, "Plan Colombia: Plan for Peace, Prosperity, and the Strengthening of the State," October 1999, http://www/ciponline.org/colombia/plancolombia.htm (accessed 4 August 2008); a later version updated for the US audience added "and to protect and promote human rights and international humanitarian law" at the end of the military strategy section and the peace process section to earlier in the document, see "Plan Colombia: Plan for Peace, Prosperity, and the Strengthening of the State," US Institute of Peace, posted 15 May 2000, http://www.usip.org/library/pa/colombia/adddoc/plan_colombia_101999.html (accessed 18 May 2009).

46. Serafino, "Colombia: Conditions and US Policy Options," 18.

47. Republic of Colombia, "Plan Colombia: Plan for Peace, Prosperity, and the Strengthening of the State." A later version updated for the US audience did not list the specific priorities for security forces in a roles and missions section. See US Institute of Peace version.

48. Republic of Colombia, "Plan Colombia: Plan for Peace, Prosperity, and the Strengthening of the State."

49. I use "Plan Colombia" to represent the US counterdrug package that served as the US component of Colombia's broader Plan Colombia. For a review of this process, see Serafino, "Colombia: US Assistance and Current Legislation." With the passage of "Plan Colombia" funding, Colombia moved from a distant third-largest recipient of US aid to a close third behind Israel and Egypt.

50. Brian E. Sheridan, "Statement before House Armed Services Committee," 23 March 2000, http://armedservices.house.gov/comdocs/testimony/106thcongress/00-03-23sheridan.htm (accessed 15 October 2008).

51. Charles E. Wilhelm, "Statement before the Senate Armed Services Committee," 7 March 2000, http://www.armed-services.senate.gov/statemnt/2000/000307cw.pdf (accessed 1 October 2008).

52. David Passage, *The United States and Colombia: Untying the Gordian Knot* (Carlisle Barracks, PA: Strategic Studies Institute, March 2000), 7–11, 27, 29.

53. Passage, *The United States and Colombia: Untying the Gordian Knot,* 13–19.

54. Tod Robberson, "Colombia Plans Offensive," *Salt Lake Tribune,* 4 December 1999, http://proquest.umi.com/pqdweb?index=5&did=46862077&SrchMode=2&sid=1&Fmt=3&VInst=PROD&VType=PQD&RQT=309&VName=PQD&TS=1242675227&clientId=5904 (accessed 5 November 2008); Notes from discussions with American military officials.

55. Marks, *Colombian Army Adaptation to FARC Insurgency,* 13–15; Notes from discussions with Colombian and American military officials.

56. Robert W. Jones Jr., "Special Forces in Larandia: ODA 753 and the CERTE," *Veritas, Journal of Army Special Operations History* 2, no. 4 (2006): 85–90.

57. "Colombia: Further Details of FARC Attacks on Vigia del Fuerte and Elsewhere," *BBC Monitoring Americas—Political,* 28 March 2000, http://proquest.umi.com/pqdweb?RQT=403&TS=1237824812&clientId=5904&D

BId=G5&xsq=Colombia&xFO=CITABS&xsq1=military&xFO1=CITABS
&xOP1=AND&saved=1 (accessed 19 May 2009).

58. Larry Rohter, "Massacre in Colombia Village Reverberates: Paramilitary Terrorism Raises Questions on US Aid for Drug War," *Pittsburg Post—Gazette*, 16 July 2000, http://proquest.umi.com/pqdweb?index=2&did=56460309&Srch Mode=2&sid=1&Fmt=3&VInst=PROD&VType=PQD&RQT=309&VName=P QD&TS=1242767033&clientId=5904 (accessed 19 May 2009).

59. US Department of State, "Colombia Country Report on Human Rights Practices—2000" (Washington, DC: US DOS, 23 February 2001), http://www. state.gov/g/drl/rls/hrrpt/2000/wha/741.htm (accessed 16 March 2009).

60. Notes from discussions with American military officials; "Colombia: Armed Forces Carry Out 'Unprecedented' Anti-Drug Operation in South," *BBC Monitoring Americas—Political*, 13 July 2000, http://proquest.umi.com/pqdweb ?index=1&did=56418519&SrchMode=2&sid=1&Fmt=3&VInst=PROD&VTyp e=PQD&RQT=309&VName=PQD&TS=1242830120&clientId=5904 (accessed 4 November 2008).

61. US General Accounting Office, *Drug Control: US Assistance to Colombia Will Take Years to Produce Results* (Washington, DC: GAO, October 2000), 16.

62. Clifford Krauss, "Attacks by Colombian Rebels Appear a Response to US Plan," *New York Times*, 20 July 2000, http://proquest.umi.com/pqdweb?ind ex=0&did=56669258&SrchMode=2&sid=2&Fmt=3&VInst=PROD&VType= PQD&RQT=309&VName=PQD&TS=1242831399&clientId=5904 (accessed 19 May 2009).

63. White House, "Report on US Policy and Strategy Regarding Counterdrug Assistance to Colombia and Neighboring Countries," 26 October 2000, Center for International Policy's Colombia Program, http://ciponline.org/colombia/102601. htm (accessed 1 October 2008).

64. "Colombia: Government Dismisses 388 Officers, NCOs," *BBC Monitoring Americas—Political*, 17 October 2000, http://proquest.umi.com/ pqdweb?index=1&did=62614800&SrchMode=2&sid=1&Fmt=3&VInst=PRO D&VType=PQD&RQT=309&VName=PQD&TS=1242839474&clientId=5904 (accessed 4 November 2008).

65. Linda Robinson and Ruth Morris, "Colombia's Messy, Complicated War," *US News and World Report*, 4 September 2000, http://web.ebscohost.com/ ehost/detail?vid=7&hid=104&sid=77ac7b0e-8678-4d6b-8bb2-49ddfb6f1153%4 0sessionmgr109&bdata=JnNpdGU9ZWhvc3QtbGl2ZQ%3d%3d#db=mth&AN =3483857 (accessed 20 May 2009).

66. White House, "Report on US Policy and Strategy Regarding Counterdrug Assistance to Colombia."

67. Brian E. Sheridan, "Statement before House Committee on International Relations Subcommittee on Western Hemisphere," 21 September 2000.

68. GAO, *Drug Control: US Assistance to Colombia Will Take Years to Produce Results*, 19–22, 26.

69. US General Accounting Office, *Drug Control: Challenges in Implementing Plan Colombia* (Washington, DC: GAO, 12 October 2000), 4–8.

70. GAO, *Drug Control: Challenges in Implementing Plan Colombia*, 10–12; GAO, *Drug Control: US Assistance to Colombia Will Take Years to Produce Results*, 22–25. For more on helicopter deliveries, see Serafino, "Colombia: US Assistance and Current Legislation," 21. Serafino provided the following counterdrug brigade helicopter deliveries: 33 UH-1N—18 in October 2000 and 15 on 2 February 2001; 14 UH-60—3 in July 2001 and remainder by end of 2001; and 30 UH-II—3 scheduled in December 2001 and remainder by May 2002; Notes from discussions with American military officials. The captain became a member of the escort team for US visitors to the counterdrug battalions and the Colombian Army commander would refer to him in their presence as his very best captain who had not been good enough for the Americans.

71. Juan Forero, "Colombia Says Rebels Have Killed 56 Troops," *New York Times*, 21 October 2000, http://proquest.umi.com/pqdweb?index=39&did=62797057&SrchMode=1&sid=4&Fmt=3&VInst=PROD&VType=PQD&RQT=309&VName=PQD&TS=1242938270&clientId=5904 (accessed 5 November 2008); Marks, *Colombian Army Adaptation to FARC Insurgency*, 16–17; Angel Rabasa and Peter Chalk, *Colombian Labyrinth: The Synergy of Drugs and Insurgency and Its Implications for Regional Stability* (Santa Monica, CA: Rand, 2001), 44–45.

72. Juan Forero, "Colombian Army Goes High Up to Fight Rebels," *New York Times*, 19 December 2000, http://proquest.umi.com/pqdweb?index=0&did=65287103&SrchMode=2&sid=1&Fmt=3&VInst=PROD&VType=PQD&RQT=309&VName=PQD&TS=1242942108&clientId=5904 (accessed 4 November 2008); Colombian Army, *Colombian Army Military History* (Bogotá: E3 Section, Army Historical Studies Center, 2007) 342.

73. Mario Montoya, "Ejercito Nacional: Informacion del Comandante," Briefing, May 2006.

74. Brian M. Jenkins, "Colombia: Crossing a Dangerous Threshold," *The National Interest* (Winter 2000/2001): 48, 51–52, 54–55.

75. "Colombia: New Commanders of Divisions Attend Military 'Summit,'" *BBC Monitoring Americas—Political*, 18 January 2001, http://proquest.umi.com/pqdweb?index=2&did=66907118&SrchMode=2&sid=1&Fmt=3&VInst=PROD&VType=PQD&RQT=309&VName=PQD&TS=1243026063&clientId=5904 (accessed 4 November 2008).

76. "Colombia: Military 'Summit' Analyses Tasks; Four Mobile Brigades Being Set Up," *BBC Monitoring Americas—Political*, 27 January 2001, http://proquest.umi.com/pqdweb?index=0&did=67351439&SrchMode= 2&sid=2&Fmt=3&VInst=PROD&VType=PQD&RQT=309&VName=PQD&TS=1243027271&clientId=5904 (accessed 4 November 2008).

77. Marks, *Colombian Army Adaptation to FARC Insurgency*, 15–21; Thomas A. Marks, "Colombian Army Counterinsurgency," *Crime, Law & Social Change* (July 2003): 100.

78. Jorge Enrique Mora, *Guia de Planeamiento Estrategico, 2001*. Fuerzas Militares de Colombia Ejercito Nacional, 2001.

79. Marks, *Colombian Army Adaptation to FARC Insurgency*, 31–32.

80. Marks, "Colombian Army Counterinsurgency," 99.

81. US Department of State, "Colombia Country Report on Human Rights Practices—2000" (Washington, DC: US DOS, 23 February 2001), http://www.state.gov/g/drl/rls/hrrpt/2000/wha/741.htm (accessed 16 March 2009); Peter Pace, "Statement before the House Appropriations Defense Subcommittee," 4 April 2001, http://www.armed-services.senate.gov/statemnt/2001/011025pace.pdf (accessed 8 March 2009); Serafino, "Colombia: US Assistance and Current Legislation," 21.

82. Marks, *Colombian Army Adaptation to FARC Insurgency*, 25.

83. Paul de la Garza and David Adams, "Military Aid . . . From the Public Sector," *St. Petersburg Times*, 3 December 2000, http://proquest.umi.com/pqdweb?index=3&did=64917780&SrchMode=2&sid=1&Fmt=3&VInst=PROD&VType=PQD&RQT=309&VName=PQD&TS=1243349412&clientId=5904 (accessed 4 March 2009).

84. David Adams and Paul de la Garza, "Contract's End Hints of Colombia Trouble," *St. Petersburg Times*, 13 May 2001, http://www.sptimes.com/News/051301/news_pf/Worldandnation/Contract_s_end_hints_.shtml (accessed 4 March 2009); Jason Vauters and Michael L.R. Smith. "A Question of Escalation—From Counternarcotics to Counterterrorism: Analyzing US Strategy in Colombia," *Small Wars and Insurgencies* (June 2006): 181; Notes from discussions with Colombian and American military officials.

85. Stephen Johnson, "Helping Colombia Fix Its Plan to Curb Drug Trafficking, Violence, and Insurgency," *The Heritage Foundation Backgrounder*, No. 1435, 26 April 2001, 8, http://www.heritage.org/Research/LatinAmerica/bg1887.cfm (accessed 15 September 2008).

86. Juan Forero, "Rebel Force in Colombia Repatriates 242 More P.O.W.'s," *New York Times*, 29 June 2001, http://proquest.umi.com/pqdweb?index=2&did=74837004&SrchMode=2&sid=1&Fmt=3&VInst=PROD&VType=PQD&RQT=309&VName=PQD&TS=1243371882&clientId=5904 (accessed 4 November 2008).

87. US Department of State, "Colombia Country Report on Human Rights Practices—2001" (Washington, DC: US DOS, 4 March 2002), http://www.state.gov/g/drl/rls/hrrpt/2001/wha/8326.htm (accessed 16 March 2009).

88. Max G. Manwaring, *Non-State Actors in Colombia: Threats to the State and to the Hemisphere* (Carlisle Barracks, PA: Strategic Studies Institute, May 2002), 6; Human Rights Watch, *The "Sixth Division": Military-paramilitary Ties and US Policy in Colombia* (New York, NY: Human Rights Watch, September 2001), http://www.colombianatverket.se/files/thesixthdivision_hrw.pdf (accessed 8 March 2009).

89. Scott Wilson, "Colombian General Convicted in Killings; Collaboration with Paramilitaries Seen," *Washington Post*, 14 February 2001, http://proquest.umi.com/pqdweb?index=12&did=68578862&SrchMode=2&sid=1&Fmt=

3&VInst=PROD&VType=PQD&RQT=309&VName=PQD&TS=124336
4549&clientId=5904 (accessed 4 November 2008).

90. Scott Wilson, "Colombia's Other Army; Growing Paramilitary Force Wields Power With Brutality," *Washington Post*, 12 March 2001, http://proquest.umi.com/pqdweb?index=6&did=69550386&SrchMode=2&sid=2&Fmt=3&VInst=PROD&VType=PQD&RQT=309&VName=PQD&TS=1243365516&clientId=5904 (accessed 4 November 2008).

91. US Department of State, "Colombia Country Report on Human Rights Practices—2001" (Washington, DC: US DOS, 4 March 2002), http://www.state.gov/g/drl/rls/hrrpt/2001/wha/8326.htm (accessed 16 March 2009); Michael Radu, "E-Notes Colombia: A Trip Report," Foreign Policy Research Institute, 1 June 2001, http://www.fpri.org/enotes/20010601.radu.colombiatrip.html (accessed 6 August 2008).

92. Scott Wilson, "War With an Absent Army; In Contested Region, Colombian Government Finds Some Towns Too Dangerous to Protect," *Washington Post*, 3 August 2001, http://proquest.umi.com/pqdweb?index=1&did=76989264&SrchMode=2&sid=3&Fmt=3&VInst=PROD&VType=PQD&RQT=309&VName=PQD&TS=1243367018&clientId=5904 (accessed 4 November 2008).

93. Juan Forero, "Colombia's Army Rebuilds and Challenges Rebels," *New York Times*, 2 September 2001, http://proquest.umi.com/pqdweb?index=0&did=79411237&SrchMode=2&sid=1&Fmt=3&VInst=PROD&VType=PQD&RQT=309&VName=PQD&TS=1243439417&clientId=5904 (accessed 4 November 2008).

94. John A. Cope, "Colombia's War: Toward a New Strategy" (Washington, DC: National Defense University, Institute for National Strategic Studies, Strategic Forum 194 (October 2002)), 5.

95. US DOS, "Colombia Country Report on Human Rights Practices—2001"; Scott Wilson, "Colombia Increases Military's Powers; Law Could Threaten US Aid Disbursement," *Washington Post*, 17 August 2001, http://proquest.umi.com/pqdweb?index=15&did=77942545&SrchMode=2&sid=1&Fmt=3&VInst=PROD&VType=PQD&RQT=309&VName=PQD&TS=1243445918&clientId=5904 (accessed 27 May 2009).

96. K. Larry Storrs and Nina M. Serafino. "Andean Regional Initiative (ARI): FY2002 Assistance for Colombia and Neighbors," Congressional Research Service Report, updated 21 December 2001, 13.

97. Human Rights Watch, *The "Sixth Division": Military-paramilitary Ties and US Policy in Colombia*, 96.

98. Marks, *Colombian Army Adaptation to FARC Insurgency*, 26.

99. US Department of State, "International Narcotics Control Strategy Report for 2001," Colombia Section (Washington, DC: US DOS, 1 March 2002), http://www.state.gov/p/inl/rls/nrcrpt/2001/rpt/8477.htm (accessed 1 October 2008).

100. Juan Forero, "Colombian Troops Move on Rebel Zone as Talks Fail," *New York Times*, 11 January 2002, http://proquest.umi.com/pqdweb?index=3&d

id=99067316&SrchMode=2&sid=6&Fmt=3&VInst=PROD&VType=PQD&RQ
T=309&VName=PQD&TS=1243525371&clientId=5904 (accessed 4 November
2008).

101. "Colombia: Army Commander Says 13,000 Troops Advancing Toward
Demilitarized Zone," *BBC Monitoring Americas—Political*, 14 January 2002,
http://proquest.umi.com/pqdweb?index=1&did=99540960&SrchMode=2&sid=
1&Fmt=3&VInst=PROD&VType=PQD&RQT=309&VName=PQD&TS=1243
527326&clientId=5904 (accessed 4 November 2008).

102. Montoya, "Ejercito Nacional: Informacion del Comandante," Briefing;
US DOS, "Colombia Country Report on Human Rights Practices—2001."

103. Colombian Army, "Disposicion Numero 000002 Por medio de la
cual se reorganiza el Ejercito Nacional," 4 February 2002. Document signed by
Colombian Army Commanding General Jorge Enrique Mora. This document
reorganized the Army in accordance with Article 29 of Law 1512 of 2000.

104. "Colombian Military Hold Emergency Meeting to Debate
Counterterrorism Strategy," *BBB Monitoring Americas—Political*, 13 February
2002, http://proquest.umi.com/pqdweb?index=0&did=107302002&SrchMode=
2&sid=5&Fmt=3&VInst=PROD&VType=PQD&RQT=309&VName=PQD&T
S=1243525103&clientId=5904 (accessed 4 November 2008).

105. Scott Wilson, "Colombian Army Ordered Into Haven As Rebel Talks
End," *Washington Post,* 21 February 2002, http://proquest.umi.com/pqdweb?in
dex=5&did=109434508&SrchMode=2&sid=1&Fmt=3&VInst=PROD&VType
=PQD&RQT=309&VName=PQD&TS=1243529560&clientId=5904 (accessed
4 November 2008).

106. Notes from discussions with senior Colombian Military officials; for air
operations, see "Colombian Air Force Chief: "Several Guerrilla Leaders Targeted,"
BBC Monitoring Americas—Political, 4 March 2002, http://proquest.umi.com/
pqdweb?index=1&did=110023051&SrchMode=2&sid=4&Fmt=3&VInst=PRO
D&VType=PQD&RQT=309&VName=PQD&TS=1243540735&clientId=5904
(accessed 4 November 2008); for units, see Colombian Army, *Colombian Army
Military History*, 346–350, which refers to the operation as Operation Tierra
de Honor (TH); Scott Wilson, "Colombians Ill-Prepared for Prolonged War on
Rebels," *Washington Post*, 3 March 2002, http://proquest.umi.com/pqdweb?ind
ex=7&did=110002917&SrchMode=2&sid=1&Fmt=3&VInst=PROD&VType=
PQD&RQT=309&VName=PQD&TS=1243536257&clientId=5904 (accessed
4 November 2008).

107. "Colombia: Commander Says Military to Ask for More Money," *BBC
Monitoring Americas—Political*, 26 March 2002, http://proquest.umi.com/pqd
web?index=0&did=111527883&SrchMode=2&sid=3&Fmt=3&VInst=PROD
&VType=PQD&RQT=309&VName=PQD&TS=1243536700&clientId=5904
(accessed 4 November 2008).

108. Andrew Selsky, "Colombia Leader Gives His Military More Power,"
San Antonio Express—News, 1 March 2002, http://proquest.umi.com/pqdweb?i
ndex=3&did=1171854151&SrchMode=2&sid=3&Fmt=3&VInst=PROD&VTyp

e=PQD&RQT=309&VName=PQD&TS=1243540085&clientId=5904 (accessed 28 May 2009).

109. "Colombia: Investigations Show FARC Controlled Drug Crops, Trafficking in ex-DMZ," *BBC Monitoring Americas—Political*, 15 March 2002, http://proquest.umi.com/pqdweb?index=0&did=110486053&SrchMode=2&sid= 8&Fmt=3&VInst=PROD&VType=PQD&RQT=309&VName=PQD&TS=1243 542032&clientId=5904 (accessed 4 November 2008).

110. Rodman, Peter. "Testimony before House International Relations Committee Subcommittee for Western Hemisphere," 11 April 2002, http://ciponline.org/colombia/02041104.htm (accessed 15 October 2008).

111. Wilson, "Colombians Ill-Prepared for Prolonged War on Rebels."

112. Marc Grossman, "Testimony of Ambassador Marc Grossman, Under Secretary of State for Political Affairs before the House Appropriations Committee's Subcommittee on Foreign Operations on US Assistance to Colombia and the Andean Region," 10 April 2002, http://www.ciponline.org/colombia/02041001. htm (accessed 29 May 2009).

113. Nina M. Serafino, "Colombia: Summary and Tables on US Assistance, FY1989–FY2003," Congressional Research Service Report, 3 May 2002, 4.

114. Gary D. Speer, "Posture Statement of Major General Gary D. Speer, United States Army Acting Commander in Chief United States Southern Command before 107th Congress Senate Armed Services Committee," 5 March 2002, http://www.ciponoline.org/colombia/02030501.htm (accessed 1 October 2008).

115. Scott Wilson, "No Sanctuary from Colombian War; Army Was Absent During Massacre at Village Church," *Washington Post*, 9 May 2002, http://proquest.umi.com/pqdweb?index=5&did=118615149&SrchMode=2&sid=1& Fmt=3&VInst=PROD&VType=PQD&RQT=309&VName=PQD&TS=124362 5034&clientId=5904 (accessed 28 May 2009).

116. "Colombian Military: Pastrana Counterterrorism Bid Bound to Fail," *BBC Monitoring Americas—Political*, 6 May 2002, http://proquest.umi.com/pqdweb?index=0&did=117885474&SrchMode=2&sid=2&Fmt=3&VInst=PRO D&VType=PQD&RQT=309&VName=PQD&TS=1243626112&clientId=5904 (accessed 4 November 2008).

117. "Colombia's Pastrana Beefs Up Military, Offers Reward for Rebel Leaders' Arrest," *BBC Monitoring Americas—Political*, 29 June 2002, http://proquest.umi.com/pqdweb?index=0&did=130362711&SrchMode=2&sid=1& Fmt=3&VInst=PROD&VType=PQD&RQT=309&VName=PQD&TS=124369 3667&clientId=5904 (accessed 4 November 2008).

118. "Colombia: Pastrana Praises Armed Forces, Police for Their Anti-Rebel Efforts," *BBC Monitoring Americas—Political*, 22 July 2002, http://proquest.umi.com/pqdweb?index=0&did=140519251&SrchMode=2&sid=1& Fmt=3&VInst=PROD&VType=PQD&RQT=309&VName=PQD&TS=124369 7101&clientId=5904 (accessed 4 November 2008).

119. Nina M. Serafino, "Colombia: The Uribe Administration and Congressional Concerns," Congressional Research Service Report, 14 June 2002.

Chapter 3

"Restoring Order and Security"[1]

The Uribe Presidency: Democratic Security and Consolidation (2002–2008)

Security is not achieved simply through the efforts of the Armed Forces and the National Police. This is an effort of the entire State and all Colombians. A strong state structure, supported by citizen solidarity, guarantees the rule of law and the respect of rights and civil liberties.

President Alvaro Uribe[2]

. . . Congress gave us Expanded Authority to use counter-drug funds for counter-terrorism missions in Colombia because it concluded that there is no useful distinction between a narcotrafficker and his terrorist activity, hence the term narcoterrorist. This link between narcotics trafficking and terrorism in Colombia was . . . recognized in National Security Presidential Directive 18 (NSPD-18) concerning support to Colombia. Operations today are more efficient and effective because our expanded authorities allow the same assets to be used to confront the common enemy found at the nexus between drugs and terror. Expanded Authority permits greater intelligence sharing and allows Colombia to use US counterdrug funded equipment for counterterrorism missions. Expanded Authority from Congress is essential to this command's ability to deal with narcoterrorists.

General James T. Hill, USSOUTHCOM Commander[3]

On 7 August 2002, in the midst of an attempted Revolutionary Armed Forces of Colombia (FARC) mortar attack on the Presidential palace that reportedly killed and wounded over a hundred in Bogotá, 50-year old Alvaro Uribe became President of Colombia. Having served in numerous appointed and elected positions—including mayor of Medellín, governor of Antioquia, and a national senator for two 4-year terms—Uribe's "Firm Hand, Big Heart" campaign captured his reputation as a law-and-order advocate and a social reformer. His 100-point campaign program included security measures—increasing the Ministry of National Defense

95

(MOD) budget, doubling the size of the military, and creating a million-man militia to assist the security forces—as well as political, social, and economic proposals. Uribe had established a reputation as an honest, charismatic, and fiscally skillful politician with ties as governor to Community Associates of Rural Vigilance (CONVIVIR) forces.[4] He would prove to be a hard-working, tireless micromanager who refused to accept substandard performance.[5] Given the nationwide guerrilla attacks in early August and the lack of an adequate police presence in almost 50 percent of the municipalities—no presence in 184 and a presence "so weak as to be virtually nonexistent" in another 370—questions existed about Uribe's ability to turn promises into actions as an independent without the support of a political party, how he would interact with his security forces, and the capability of the security forces to turn increased resources into improved performance against the ongoing, widespread guerrilla attacks.[6]

Uribe Takes Charge: 2002

Within days, Uribe responded to the security situation. First, he established his security team. The first female Minister of Defense, Marta Lucia Ramirez, led the MOD with the Army commander, General Jorge Enrique Mora replacing General Fernando Tapias as the Colombian Military (COLMIL) commander, with General Carlos A. Ospina assigned as the Army commander, and with Major General Teodoro Campo Gomez promoted to Colombian National Police (CNP) director. The Navy and Air Force commanders remained unchanged. Then, to open the roads to traffic, on 8 August in the department of Cesar, Uribe called on Colombians to support their security forces and launched his "one-million-member" Citizen Security Plan by creating an initial group of 600 unarmed volunteers equipped with radios to provide information to the security forces.[7] Finally, after over 100 Colombians had been killed in 4 days, on 12 August Uribe declared a state of emergency that established a one-time 1.2 percent emergency tax for individuals and companies with $60,000 in assets.[8] This raised $800 million in taxes to finance 6,000 Army professionals for two additional mobile infantry brigades (BRIMs), 10,000 additional police officers—primarily rural police or *Carabineros*, administration of the civilian information or informant network, and reduce the military budget deficit.[9] In September, Uribe used his state-of-emergency decree to establish two "Rehabilitation and Consolidation Zones." In each zone, military commanders controlled all security forces and could conduct searches, question civilians, impose curfews, and restrict travel.[10] The zone in the northeast included parts of the departments of Arauca and Sucre. The former *zona de despeje* constituted the second zone.[11]

In addition to national support, additional resources, and new powers, Uribe brought a sense of urgency and greater civilian involvement in security matters. An experienced government administrator, Minister Ramirez stated in late August, "As we are in the process of defining what type of ministry we need we shall also have to make decisions about what kind of officials have the best characteristics for this ministry."[12] In late August after Uribe publically criticized Army officers in one department for their slow response to guerrilla activities, the brigade commander responded: "At times we do not have enough soldiers" but his forces were "working hard throughout the region. This situation is not new. It is many years old." Ramirez characterized Uribe's comments as not a reprimand but as a call for the security forces to become more proactive, more responsive, and less defensive—"a recommendation for all of us to be on the alert. . . We must anticipate events, act wisely but quickly and not be simply on the defensive. If we already know that there are places where there are guerrillas, we need the security forces to be present there to prevent regrettable incidents."[13] To make this possible, Uribe supported the COLMIL and CNP expansion plans—professionals, BRIMs, mountain battalions, a commando battalion, rural police squadrons, intelligence fusion, infrastructure security battalions—but not their timelines. Uribe wanted more, he wanted it faster, and in the case of road security and local security forces, he wanted something new. The CNP's proposed 4- to 5-year timeline to reoccupy municipalities with no police presence became 18 months.[14] In addition to its counterdrug support of the Antinarcotics Police Directorate (DIRAN)—Air Service and three 166-man *Junglas* companies—the Narcotics Affairs Section (NAS) agreed to assist the CNP by training and equipping over 60 150-man *Carabinero* or rural police squadrons armed similar to Army units and by training and equipping policemen for the unoccupied municipalities.[15] In October, Ramirez announced MOD goals for the police to reoccupy 120 major municipalities in 6 months and for the security forces to expand by 100,000 personnel—55,000 in 6 months and another 45,000 by December 2003.[16] Once raised, these forces became the first installment of a multiyear program known as Plan *Choque* or Shock Plan. Uribe's work ethic and his demand for results, in the words of a former Military Group (MILGP) commander, "cascaded down to every rank in the Police and Military."[17]

American officials seemed inclined to support Uribe's efforts, but they sought evidence of greater Colombian commitment to addressing its security problems. Meeting with 800 Colombian business leaders in late July, Ambassador Anne W. Patterson echoed many of Uribe's proposals—additional security forces, tax increases, legal reform—and stressed,

"Priority must be to restore security." She added, "This means that the United States is ready to invest more if the Colombians invest more in their own security."[18] On 2 August, President George W. Bush signed the supplemental funding law that provided for the initial training of 18 Brigade responsible for infrastructure or oil pipeline security in the Arauca department—the first major US noncounterdrug-related program. In addition, the law provided expanded authorities that authorized "a unified campaign against narcotics trafficking, against activities by organizations designated as terrorist organizations and . . . actions to protect human health and welfare in emergency circumstances, including undertaking rescue operations." With this change from a strictly counternarcotics focus, the United States and Colombian views of the Colombian security challenges began to align. In late November, the Fiscal Year 2003 Intelligence Authorization Act permitted intelligence sharing and provided intelligence-related funds to support the unified campaign in Colombia.[19] Earlier that month, Bush issued NSPD-18, Supporting Democracy in Colombia, which provided guidance for the new US policy in Colombia.[20]

As American policy changed, USSOUTHCOM and the MILGP broadened their former counterdrug approach with the MOD and the COLMIL. Problems immediately identifiable by the Americans—some the result of previous US efforts—included no strategic plan, poor unity of effort, limited sharing of only counterdrug intelligence, Joint Task Force–South (JTF-S) and BRCNA deployment restrictions, priority for support and repair parts, and five separate helicopter fleets. Rebuilding the confidence of senior COLMIL leaders became an important task—particularly given the need for increased Colombian resources, for long-term US support, and for continued human rights vetting.[21] In the last months of the Pastrana administration, the MILGP had continued its counterdrug training and had worked with COLMIL on planning for a second counternarcotics brigade, the reorganization of the four-battalion Special Forces (SF) Brigade to a three-battalion brigade to provide for a commando battalion to attack high-value targets (HVTs), the reorganization of the BRCNA, and the training for an infrastructure security unit—18 Brigade. An eight-man Joint Planning and Assistance Team (JPAT) working much of this effort included five reservists or National Guardsmen on temporary duty. In August, the MILGP reviewed the new authorities with the COLMIL and stressed the importance of human rights vetting of Army units to gain access to US assistance. In a September briefing on the counterdrug brigade reorganization and infrastructure security, Ambassador Patterson and General Mora agreed that a smaller, reorganized, 2,100-man, all-professional BRCNA would be placed under Army control for use where needed. Near the end of the year, General James T. Hill, USSOUTHCOM commander since July, and Mora met to discuss a US concept to place American military personnel, a Planning Assistance Training Team (PATT), in selected Army units.[22] In this give-and-take process after the US policy change, neither side accepted all the suggestions made and those that were accepted required planning, time, and resources to become a reality. For example, the training for 18 Brigade and the commando battalion, discussed for months, did not begin until October and continued into 2003.

Uribe sought to reform the MOD at the same time as he expanded and improved his security forces. A USSOUTHCOM official emphasized the Colombian need for a "good national military strategy—a game plan to go on—to prosecute."[23] At the end of August, a retired US Air Force major general arrived to assist the MOD in the development of a national military strategy. Working with senior MOD civilian and military personnel, he played an advisory role in the MOD reorganization and in the creation of a national military strategy. Opening an October strategy seminar, the COLMIL commander emphasized his support of civilian control of the military, jointness, and better cooperation and planning within the MOD—all concepts emphasized by the Americans.[24] What he may have actually meant was control by the President, jointness in effect but not in structure given the absence of a Colombian Goldwater-Nichols Act, and better cooperation and planning in support of the major combatant—the Colombian Army. By December, the MOD had reorganized with two vice-ministers: one for Budget and one for Policies and Foreign Affairs. Uribe provided clear guidance: change the MOD by taking things that work and improving them without asking for new legislation, additional funding, or new organizations. All his initial priorities—the security of the roads, reduction of kidnappings, and expansion of the civilian cooperative (informant) program—required working within each military service, between the military services, and between governmental agencies.[25] All of these actions, and those in other governmental ministries, generated the programs for what would become Uribe's Democratic Security and Defense Policy (DSDP).

The FARC and the National Liberation Army (ELN) continued guerrilla attacks throughout the country attempting to discredit and paralyze the government by damaging critical infrastructure—particularly electrical grids, communications systems, and oil pipelines; disrupting road traffic and commerce between urban areas; and terrorizing the population through bombings, killings, and forced displacements. Narcoterrorist or illegal armed group strength at the end of 2002 totaled over 36,000—17,000 FARC, less than 4,000 ELN, and over 12,000 United Self-Defense Groups of Colombia (AUC).[26] As in 2001, AUC strength continued to grow faster in 2002 than either the FARC or the ELN. In December, the AUC agreed to a unilateral ceasefire as a precondition to negotiate conditions for demobilization with the government. To confront these threats, the security forces had grown to 313,000 personnel by the end of the year—203,000 military with over 166,000 soldiers and 110,000 police—but the narcoterrorists maintained the initiative.[27]

Democratic Security and Defense Policy and Plan *Patriota*: 2003

In early 2003, COLMIL published its first national military strategy—a document developed in parallel with and in support of the government's Democratic Security Policy that would be published in June. The strategic objectives called for improving the operational capability of each service, destroying the will of the narcoterrorists by rapid and decisive operational successes and protecting the population, strengthening legitimacy through transparency and respect for human rights, maintaining deterrence against external threats, and improving the proficiency of military personnel. The new operational concept called for rapid reaction based on improved collection and sharing of intelligence, integrated communications, and responsive air support for ground operations. To execute this concept, the strategy addressed six lines of action. First, the 4-year, phased expansion plan—Plan *Choque*—called for additional units: mobile brigades, mountain battalions, antiterrorism units (AFEUR), antikidnapping and extortion units (GAULA), naval riverine forces, and 40-man platoons of *soldados campesinos* for local security in towns. Other lines of action included increasing the size and capabilities of each service; improving technical, human, and strategic intelligence; increasing antiterrorism units; improving joint air support capabilities; and strengthening integrated action.[28] The Colombian Army issued operational guidance supporting the COLMIL strategy.[29] These documents provided a long-term framework that underlay day-to-day decisions and improved military capabilities as the security forces struggled with the narcoterrorists.

The COLMIL Infrastructure Security (ISS) Plan had three phases: I—security of the first 110 miles of the Cano Limon-Covenas Oil Pipeline in the department of Arauca; II—security of the remaining 367 miles of the pipeline between Arauca and the Caribbean coast; and, III—security of 338 critical economic infrastructure sites. United States support for phase I provided the Colombian Army its largest and its first noncounterdrug assistance package. This 2-year, $99 million program—scheduled to begin in January 2003 and end in December 2004—provided $71 million for 10 helicopters to be delivered by May 2004, $15.4 million for training, and \$12.7 million for equipping 18 Brigade.[30] In January, US 7 SF Group soldiers arrived to conduct their training missions—70 Americans worked with 18 Brigade in Arauca, 25 with the BRCNA at Larandia, and 15 with the commando battalion at the Tolemaida Training Center.[31] Designated part of a "Rehabilitation and Consolidation" zone, Arauca—roughly the size of New Hampshire—had three generations of inhabitants who knew only the ELN and recently the *autodefensas*. An American described the situation there as trying to train-the-trainer in the middle of a gunfight.[32] Reflecting the view of many Colombians, another American observed, "The paramilitaries are bad guys, but they're *good* bad guys." Still, what the Americans found impossible to comprehend was "the designation of a war zone as a civilian 'crime scene.'" What appeared to be a "legal fiction" to them remained a legal reality for the Colombian security forces who operated under peacetime law.[33] Problems with the American timelines proved another Colombian reality. As with the previous US priority program—training and equipping the counterdrug BRCNA—the ISS training had to be adjusted to accommodate equipment delays caused by late decisions, by the US procurement system, and now by the competing demands in Afghanistan and Iraq. The helicopter delivery date shifted from May 2004 to April 2005 to June and had not been completed by September, over a year later than planned. Much of the programmed equipment—weapons, ammunition, night vision goggles, helmets, medical supplies—arrived late as delivery lead times approached 2 years. A partial shipment of night vision goggles ordered in late 2003 arrived in June 2005. Despite these delays, the security situation improved as attacks on the pipeline dropped from 170 in 2001 to 41 in 2002 to 34 in 2003 and arrests soared from 3 between 1986 to 2001 to almost 600 in 3 years.[34] As usual, the narcoterrorists adjusted by shifting their attacks further down the pipeline away from Arauca and by changing their attacks from the pipeline to the electrical system that pumped the oil. Through these delays, Colombian security officials continued to find American bureaucratic procedures cumbersome, unresponsive to their needs, and frustrating.

Despite the expanded authorities of the American officials and the efforts of the Uribe administration and its security forces, neither the US emphasis on human rights nor the level of violence diminished. In January, Embassy officials decertified 1 Air Combat Command (CACOM)—the first time a Colombian Air Force unit lost US assistance. Disagreement between the Embassy and Air Force over the details of the bombing of the town of Santo Domingo in 1998 that caused civilian casualties continued and eventually led to the resignation of the Air Force commander in August.[35] In February, as it had shortly after the approval of the expanded authorities the previous year, the MILGP again asked the Army to initiate vetting for additional military units.[36] The 7 February FARC car-bomb attack on the El Nogal nightclub in Bogotá that left 35 dead and 173 injured served as a wake-up call to many of the Colombian elite about their vulnerability to narcoterrorism. A week later, the crash of an American surveillance aircraft that left two dead, an American contractor and a Colombian soldier, and three Americans as FARC hostages served as a wake-up call to many Americans about the dangers of increased involvement. Describing the security situation, the USSOUTHCOM commander testified, "Colombians suffer daily from a level of violence and terror practicality unimaginable to us." Citing 2002 figures, General Hill noted that Colombia had 1.5 million internally displaced persons; had the highest number of terrorist attacks in the world—an average of four a day, more than all the attacks in all of the other nations of the world; had the highest homicide rate—13 times that of the United States, which made homicide the most likely cause of death in Colombia; had the highest kidnapping rate in the world with 2,900 abductions; and had remained the number one producer of cocaine in the world. Ongoing US military efforts to reduce this violence included supporting the BRCNA and the commando battalion, training the infrastructure security brigade in Arauca, and sharing intelligence with the Colombians. Because the expanded authorities for 2003 made this possible, he requested expanded authorities for 2004 for his entire area of responsibility. Despite the dismal statistics, Hill ended his comments noting Uribe's 70-percent approval rating with Colombians and the positive changes underway.[37] In a different forum, Hill emphasized that US military support continued to be training and assistance—not operational—conducted by a limited number of military personnel working with Colombian security forces who had improved their combat performance and their human rights record. Alleged human rights violations by the security forces had dropped to 2 percent.[38] During that same period, 45 percent of the FARC attacks had been against civilian, not military, targets, and the ELN and the AUC had attacked civilians in 90 percent of their actions.[39] Colombia remained a violent country.

In addition to increasing security force presence throughout the country, COLMIL pursued a "decapitation" strategy aimed at the narcoterrorist leadership. To have the capability to strike HVTs—an attempt against narcoterrorist leaders or to rescue hostages—both Pastrana and Uribe supported the establishment of a special unit and better sharing of intelligence. With input from the MILGP commander and training from US Special Operations Command South (USSOCSOUTH), the Colombian Army created a Ranger-type commando battalion and established an Army

Special Operations Command (COESE) on 17 March. The COESE had two small battalions of specially trained personnel—a commando battalion and a *Lancero* group.[40] On 5 May, 75-men from the commando battalion conducted a hostage rescue operation as part of a joint team that started with Brigadier General Mario Montoya's 4th Brigade and involved the FUDRA and the Colombian Air Force—photo reconnaissance and helicopter support. On hearing the helicopters, the FARC commander ordered the hostages killed and his unit to disperse. The rescue force landed in three groups and moved to the site making no contact with the narcoterrorists. On arrival 20 minutes or so after landing, the rescuers found nine hostages murdered, three wounded—one who later died, and one unharmed. The dead included the governor of Antioquia, a former Minister of Defense, and eight members of the Armed Forces. Even with good intelligence and specially trained forces, hostage rescue proved a high-risk operation. In a televised address to the nation, President Uribe took full responsibility for the failed rescue.[41] But not all HVT operations failed. On 18 May in an operation planned over 4 weeks and supported by signals intelligence, 650 men from the SF Brigade, accompanied by Department of Administrative Security (DAS) officials and civilian prosecutors, simultaneously arrived by helicopter in four municipalities in Caqueta that served as the base for one of the FARC's most capable units and the one that had captured the American hostages—the Teofilo Forero mobile column. This operation led to the capture of 52 FARC members, to include several senior members.[42] However, US signals monitoring had indicated the American hostages remained deep in the jungle surrounded by multiple rings of security and scattered minefields that made rescue difficult. The FARC military chief, Mono Jojoy, told his subordinates, "We will never allow the gringos to be rescued. Split them up." American officials asked that no rescue attempt be made without their prior approval.[43]

With the publication of the DSDP in June, Uribe established his long-term whole-of-government approach to address Colombia's problems. The policy rested on three concepts: a lack of personal security caused Colombia's political, social, and economic troubles; the absence of government in large areas of the country caused this lack of personal security; and all elements of the government had to end this absence by integrating the nation into a whole.[44] The DSDP strategic objectives were: consolidating government control throughout the country; protecting the populace; eliminating the illegal drug trade; protecting Colombia's borders; and achieving transparent and efficient resource management. To achieve these objectives, the government pursued six policies:

(1) Coordination of actions through a National Defense and Security Council, a Joint Intelligence Committee, a reformed MOD, and interagency support teams.

(2) Strengthened state institutions—judicial, armed forces, national police, intelligence, and finances.

(3) Consolidated control of the countryside through recovery, maintenance, and consolidation of territory; a border security plan; improved urban security; elimination of illegal drug trafficking; and dismantlement of narcoterrorist and drug finances.

(4) Protection of the rights of Colombians—for persons at risk, the internally displaced, the demobilized, child combatants; and against terrorism, kidnapping and extortion, and recruiting of children; and protection of critical infrastructure—economic and roads.

(5) Cooperation for the security of all Colombians through solidarity, cooperative networks, rewards programs, and international support.

(6) Communication of state policy and actions to the populace and the international community.[45]

No one knew how well the policy would be executed, but Uribe placed high demands on his government officials, often directly calling them or their subordinates at any time of the day or night.

The DSDP strengthened the security forces and highlighted several tasks—some more traditional than others. First, it directed the COLMIL to improve mobility through increased, reorganized, and trained manpower; to improve the quality of its members and the readiness of its equipment; to improve the gathering, processing, analysis, and coordination of intelligence; to continue its human rights and international humanitarian law training; and to use its resources to deter external threats. To impede the flow of illegal drugs out of Colombia and of illegal weapons into the country, border security became a priority. Another priority became the Government Road Security Program. In addition, the policy sought to reform military service by making all members liable for combat duty regardless of educational level and it called for an increase in the role for the recently created *soldados campesino*. Second, the DSDP directed the police to reoccupy municipalities that had no police presence, to create 62 mobile field police or *Carabinero* squadrons, to build fortified rural police stations, to strengthen the highway police, to coordinate city security plans, to increase its strength by 10,000 new policemen and by 10,000 auxiliary policemen, and to build civilian cooperation networks.[46]

In addition to fighting the narcoterrorists and to executing its expansion program, the DSDP specified COLMIL requirements for infrastructure security, local security, and road security. First, the three-phased infrastructure security program—*Plan Especial Energetico Vial* (PEEV)—demanded attention. Special PEEV units were created to address phase II and III requirements as the United States continued to support phase I with training in Arauca. Second, understanding the need for permanent local security, Uribe had pledged during his campaign to arm civilians—an idea not supported by many American officials given the CONVIVIR and *autodefensas* experience. However, the discovery of a 1940's law that allowed conscript soldiers to serve a portion of their military time at home permitted the creation of 40-man *soldados campesinos*—peasant soldier—platoons. Trained as regulars, led by a professional noncommissioned officer (NCO), and serving for 2 years, many of the soldiers referred to themselves as *soldados de mi pueblo*—soldiers of my town. These *soldados campesino* platoons formed a company in the local territorial battalion. In 2003, the COLMIL planned to train 15,000 to serve in 450 municipalities in priority areas.[47] This permanent security force, composed of trained armed locals under military control, had an immediate effect. As a mayor under a FARC death threat said, "It has bettered the peace" because "[t]hey know the land. They know the people." And most important, but not stated, they are always present.[48] Third, Uribe understood the actual and psychological importance to the populace of opening the roads to traffic. However, the police proved too weak for the task and the military resisted it. In early February, Uribe met with the Minister of Defense, the two vice ministers, and his general officers. After asking the generals their views and listening to comments about road security being police work—not a military job—and a distracter from more important tasks, Uribe told them they were correct about the roads being police work. Nevertheless, if they viewed the roads as supply routes or lines of communication, then it appeared he had a military that could not, or would not, protect its lines of communication. He then described the road as a plaza where the military and the people could interact. Uribe said that convincing the people that the roads were safe was the first task; transitioning the roads to the police would be the second. He noted that when he became President every Colombian problem had become his problem and that he needed their assistance with the roads. Instead of ordering the military to do the task, he explained the problem, showed them their role, and gained their support.[49] Plan *Meteoro* or Meteor Plan was the COLMIL program to open the roads. In addition to the security provided by territorial units and the information provided by the cooperative networks, special joint Army–highway police *Meteoro* companies patrolled the major roads—establishing security and providing a show of force. After the primary roads became secure, these units shifted to the secondary roads.[50]

The COLMIL developed a multiyear campaign plan, Plan *Patriota* or Patriot Plan, to regain control of priority areas in the country. It had three phases: (1) securing Bogotá and Cundinamarca department, (2) securing the FARC rear area in Caqueta department, and (3) securing Antioquia department. The 5 Division commander, Major General Reinaldo Castellanos, conducted phase I, Operation Liberty I, between 1 June and 31 December 2003. Reinforced by the FUDRA, the SF Brigade, and Air Force assets, 5 Division destroyed multiple FARC fronts and ended the threat to Bogotá in the largest, most complex, and most successful operation conducted by Colombian forces. These operations—that eliminated seven front leaders, six deputy leaders, and seven heads of finance—handed the FARC an estimated 4 to 5 year setback.[51] With the expansion and specialization of units under Plan *Choque*, security forces began to dominate priority areas using what some referred to as a grid system. First, military forces cleared the area of narcoterrorists. Then local security forces reoccupied the area—police units and platoons of *soldados campesenos* in the towns and better armed, more mobile *Carabinero* squadrons in the municipalities around the towns. With a permanent security presence, government activities could rebuild. This became the operational concept implemented by security force commanders.[52] In September, a Colombian Army reorganization recognized the creation of 6 Division that had replaced JTF-S in late 2002, of the Army Special Operations Command created in March, and of other specialized units. Each division had two to four brigades, urban antiterrorism SF units (AFEUR), and most had mobile brigades, extra counterguerrilla battalions, and *Meteoro* companies. In addition to the FUDRA, SF Brigade, BRCNA, and Aviation Brigade, the authorizations for the six divisions totaled 6 mobile brigades, 46 infantry battalions, 38 counterguerrilla battalions, 18 support battalions, 11 engineer battalions, 11 PEEV battalions, 9 cavalry squadrons, 9 artillery battalions, 6 mountain battalions, 15 GAULA units, 14 AFEUR units, and 7 *Meteoro* companies.[53] As the Army grew larger, it became more tailored to meet specific security tasks.

Just as support for "Plan Colombia" had required time for the US Country Team to reorganize, increase manning, and develop appropriate programs, the expanded authorities created similar challenges—particularly for the MILGP in 2002 and 2003. In addition to coordinating training for Colombian units by 7 SF Group personnel, the MILGP organized liaison sections to work with the COLMIL on information operations, psychological operations, medical, intelligence, civil affairs, and engineer issues. Because temporary duty personnel—primarily reservists and guardsmen from all services serving short tours—filled many of the positions, turbulence created continuity problems. An American joint planning assistance team manned the operations planning group that worked with COLMIL and scheduled mobile training teams—staff training, military decision making, civil affairs, casualty evacuation—for the limited number of human rights vetted units.[54] Given the small divisional staffs and the even smaller brigade staffs, the appropriateness of US staff and decision making training that relied on trained and sizable staff sections might be questioned. As an American trainer noted:

> They don't plan like we do. The division commander will brainstorm with his brigade commander and the staffs, then give a verbal order. The brigade commander will write an order which is more like our FRAGO [fragmentary order] so that the general knows the brigade commander understands his concept. Then they go with it.[55]

As a US initiative in early 2003, a four-man Planning and Assistance Training Team (PATT) joined the BRCNA. Created to assist in the planning for operations, the team contained operations, intelligence, logistics, and civil affairs personnel that served 1-year tours. Many Spanish-speaking US military personnel from the Active and Reserve Components filled the teams. The functions, locations, and size of the teams changed over time, but the maximum PATT strength seldom exceeded 50.[56] Sometimes the Americans asked what Colombians needed; but just as often, they provided them with what they thought was needed.

Despite increased support by Uribe, progress of ongoing operations, and US assistance, fundamental problems remained. In August, General Edgar Alfonso Lesmes became the Colombian Air Force commander after repeated US pressure concerning a 1998 bombing incident led to the resignation of his predecessor.[57] Although security force human rights complaints had dropped, past incidents continued to affect Colombian–US relations. A more serious issue arose in November when Minister Ramirez resigned after struggling with the COLMIL leadership. Unlike under Pastrana, the Uribe team sought MOD reform and greater civilian involvement in security matters—primarily nonoperational— that previously had been considered military issues. Summarizing the problem, Ramirez said, "There is a civilian minister of defense but not a civilian ministry of defense."[58] Determining what a civilian MOD would look like, how it would work, and clarifying basic civil-military issues had proven a challenge. Some in COLMIL equated civil control of the military as control by the President, not by a civilian minister. Uribe's style of personally dealing directly with commanders reinforced this tendency. Uribe responded to this resignation by retiring the COLMIL commander. Understanding that internal disputes would not be tolerated, the new MOD team—Minister of Defense Jorge Alberto Uribe and COLMIL General Ospina—worked together with less conflict.[59] General Martin Orlando Carreno replaced Ospina as the Colombian Army commander.[60] Like many of the problems with which Colombians wrestled, MOD reform and changing institutional culture had no quick—much less easy—solution. Change required a sustained trial-and-error effort to determine what worked best in Colombia.

By the end of 2003, things had begun to change. Security forces gained the initiative in parts of Colombia. Operation Liberty I had ended the FARC threat to isolate Bogotá. Road security improved. Following the lead of President Uribe, Ospina called on his generals to make hostage rescue a major priority.[61] The new commander promised to resign if the Army failed to kill or capture a member of the FARC Secretariat within a year.[62] The AUC began demobilization negotiations with the government and one organization demobilized that fall. The FARC had failed to disrupt the local elections in October and narcoterrorist desertions increased 80 percent. In August, the Department of State (DOS) reinstated the counterdrug Air Bridge Denial Program. Drug eradication proceeded at a record pace, and the CNP established a presence in all but 18 municipalities. Those 18 were occupied by February 2004.[63] Despite these improvements, Uribe continued to pressure his security forces for results.

Plan *Patriota* and Joint Commands: 2004–2006

Phase II of Plan *Patriota* began on 4 January 2004 with a 17,000-man assault into long-held FARC base areas in Caqueta department where the "state [had] abandoned these people years ago."[64] The area of operations later expanded to include the parts of the departments of Guaviare and Meta—or roughly an area near the former *zona de despeje*. During these operations, Major General Castellanos commanded COLMIL's first joint command: Joint Task Force (JTF) *Omega*—a division equivalent comprised of Army, Navy, and Air Force units. Formed in late December 2003, JTF *Omega* consisted of the three-BRIM FUDRA, the three-battalion SF Brigade, six additional BRIMs, Military Intelligence Region 8 (RIME 8), Army aviation, one counterdrug battalion, a logistics company, a medical evacuation company, seven Navy riverine combat elements, and an Air Force air component. These operations—coordinated with the 4 Division to the north and 6 Division to the south—disrupted the FARC strategic plan by attacking the narcoterrorist support infrastructure and by engaging forces from the Eastern and Southern blocs. During 2004 and 2005, the BRIMs assigned to JTF *Omega* had US PATT personnel assigned. US logistical support proved crucial for sustaining these combat operations. Improved medical support procedures and contract flights for rotating troops into and out of the area of operations helped sustain troop morale.[65] US military assistance, which continued to remain less than counterdrug assistance, provided JTF *Omega* the capability to conduct sustained operations in an isolated region dominated by the FARC for decades. As the operations unfolded, Minister of Defense Uribe cautioned, "Plan *Patriot[a]* is not a great military operation, it is a jigsaw puzzle . . . [that] is going to last a long time. There is not going to be one great battle." Eventual success would come from the accumulation of small accomplishments—seizure of weapons, disruption of units, defection of narcoterrorists, destruction of drug facilities, capture of leaders, establishment of local governance— over a period of years.[66]

In his annual USSOUTHCOM assessment to Congress, General Hill acknowledged the progress he had witnessed during his 23 visits to Colombia. He stressed the critical role played by President Uribe and the importance of building capable government organizations that would continue after his presidency.

> President Uribe is a unique leader who has galvanized the will of the people and motivated his Armed Forces. He has personally demonstrated that one individual can change the course of events. Without his personal leadership, energy, and dedication, I don't think the Colombians would have achieved the remarkable progress we have seen. Yet his personal charisma and drive only go so far, and he well knows it. That is why he is building the structures to sustain momentum and institutionalize success beyond that of his term and beyond that of Plan Colombia.[67]

To improve COLMIL capabilities, USSOUTHCOM and the MILGP worked closely with the MOD and the COESE, assisted in the operational and logistical support of JTF *Omega* operations, focused on establishing Navy riverine combat elements, supported the reestablishment of the DOS Air Bridge Denial Program, assisted in infrastructure security planning, provided PATTs for selected human rights vetted units, worked to develop intelligence-driven operations, improved civil affairs capacities, supported creation of a Military Penal Justice Corps, assisted in establishing a Command Sergeant Major school "to develop a robust noncommissioned officer corps," and continued to train COLMIL units. Hill described military training for human rights vetted units as "a key enabler in their fight."[68] However, the majority of the Colombian security forces never saw an American trainer. During this period, about seven SF A Teams (ODA) provided training to Colombian Army and police personnel. The number of security force units training with these teams—usually small classes to address specific skills and none larger than a battalion—remained limited and focused on the BRCNA and its helicopter unit, the COESE with its commando battalion and lancero group, urban antiterrorism (AFEUR) units, phase I infrastructure security units—18 Brigade and 5 BRIM, personnel from the FUDRA and the SF Brigade, Navy riverine units, Army aviation, CNP *Carabineros*, and DIRAN *Junglas*.[69] To meet the requirements generated by the increased tempo and scale of operations undertaken after the expanded authorities, Hill requested that the cap on in-country military personnel permitted to support "Plan Colombia" be increased from 400 Department of Defense (DOD) personnel and 400 contractors to 800 DOD personnel and 600 contractors.[70] This cap, signed into law in October, did not include DOD personnel assigned to the MILGP or those involved in search and rescue activities for the three American hostages.[71]

With expanded authorities and increased combat operations, some in the US Armed Forces saw the need for a greater American military role in Colombia—something neither requested by, nor probably acceptable to, the Colombians. In addition to a number of student papers written at the Army War College and the National Defense University calling for a more direct US involvement, the US Army South (USARSO) commander—drawing on a version of the El Salvador experience—called for a US military advisory effort, primarily Army, with "boots on the ground" as a "concrete manifestation of US resolve." Defining victory as the destruction of the FARC, the dismantling of its narcotics network, and ending the war on drugs in the United States, he believed that this "job is simply too large for the Colombian Military alone." The advisory structure recommended a two-man, Spanish-speaking advisory team at each Colombian division and brigade—a combat arms or SF major or captain and a military intelligence major or captain—supported by US colonel and lieutenant colonel advisors in the COLMIL and Army staff sections. To avoid the impression of a "less-than-firm US commitment to the war against drug trafficking and terrorism," US unit advisors were to accompany their counterparts during operations. The article concluded:

> A robust US military advisory program might not bring the Colombian war to a negotiated settlement . . . nor will it ensure an ultimate military victory for the Colombian

Military; however, it can buy time to achieve victory by preventing the destruction of Colombia's political, economic, and social infrastructure by an armed, well-organized criminal group.[72]

Instead of this more aggressive advisory model, the US maintained its assist and support approach, which was proving adequate for Uribe and his security forces to reduce the narcoterrorist threat.

President's Uribe's whole-of-government DSDP constituted another jigsaw puzzle with complex social, economic, political, and legal components. Although the establishment of security and control throughout the country remained the focus of the MOD and served as the prerequisite for implementing governmental programs, other ministries had responsibility for the reestablishment of local governance and social programs. To oversee this effort, and with MILGP support, Uribe established and led the interministerial Coordination Center for Integrated Action (CCAI). Permanently manned by personnel from the ministries, the CCAI coordinated civil affairs in newly secured areas. Initially the CCAI focused on 11 areas that encompassed 58 municipalities.[73] The CCAI worked to develop ways to maintain control by reducing crime and drug-related activities, to strengthen the justice system, to establish humanitarian and social programs, to encourage economic activity, and to reintegrate the populace into Colombia. MILGP, NAS, and the US Agency for International Development (USAID) personnel worked with the CCAI in support of consolidation efforts.[74] At the same time, US agencies continued to work their nonmilitary and counterdrug programs in support of "Plan Colombia." After a multiyear effort by the Embassy supporting judicial reform and the establishment of an accusatorial system similar to that in the United States to reduce the backlog and to speed up cases, Colombia reformed its judicial system in 2004 and began a phased implementation that would take effect for the whole country in January 2008. During the 4-year transition, the United States provided training for judges, prosecutors, and criminal investigators in the new system.[75] As US assistance continued to fund counterdrug, military, and nonmilitary programs, the Colombian government—just as its security forces—developed and refined its programs.

Despite a sharp decline in violence, some wondered about the focus and conduct of Uribe's program. Questions arose about the sustainability of military operations "for months or years on the zones that have historically been controlled by the guerrillas [and] where they have great popular support." Some attributed government progress to a "tactical retreat" by the FARC, noted the lack of success in capturing or killing key leaders, and repeated human rights concerns voiced by nongovernmental organizations (NGOs).[76] By declining to attack the security forces, dispersing into small units, and moving deeper into remote areas, the FARC maintained its forces while protecting its leadership. In February, HVT operations captured the finance officer from 14 Front and gained valuable information from her computer about FARC drugs and weapons trafficking. Earlier, Simon Trinidad, a senior FARC negotiator involved in narcotrafficking, had been arrested in Ecuador and extradited to Colombia.[77] However, no member of the FARC Secretariat had been killed or captured. Human rights groups continued their complaints, but the DOS annual human rights report acknowledged that different standards—one applying legal compliance and one using other criteria—led to "drastically divergent understandings . . . [that] deepened already profound mutual suspicions" and to under-reporting by the government and over-reporting by the human rights groups.[78] After the December 2003 AUC ceasefire, negotiations continued that produced results the following summer when the AUC signed an agreement to demobilize by the end of 2005.[79] Demobilization of the organization responsible for the majority of the human rights violations and a significant amount of violence would eliminate a serious threat during a time that the ELN—reduced in numbers—expressed interest in renewed negotiations. How the demobilized paramilitary members would be reintegrated into society and held accountable for their deeds remained to be resolved, but most saw the AUC demobilization agreement as another Uribe success that improved security.

Joint commands that integrated each of the military services under a single commander for a particular mission or a particular area of operations, normally overseas, had become a dominant US military concept after the Goldwater-Nichols Act of 1986 replaced jointness as a function with jointness as a structure. In theory, joint commands reduced redundancy and waste, fostered interoperability, eliminated service rivalry, and improved combat performance. For this to happen in the United States, it had required institutional reform driven by a legal mandate from Congress. Even then, it took years, if not decades, to develop a joint mindset, create joint doctrine, and establish joint procedures. What appeared normal to many Americans in 2004 remained a difficult, alien concept for COLMIL—just as it had been for the US military before 1986—because of history, institutional cultures, limited resources, constitutional mandates, and governmental procedures. USSOUTHCOM and the Embassy emphasized joint and interagency organizations to the MOD and the COLMIL. In the late 1990s, a US Joint Intelligence Center (JIC) had been established to share counterdrug intelligence and to model joint capabilities. The US-sponsored counterdrug

JTF-S—largely an Army and a CNP organization—had a joint staff and a naval officer had served briefly as its second commander. In December 2003, JTF *Omega* became a COLMIL joint command. Predominantly an Army organization, JTF *Omega* served as a small, long-term laboratory for evaluating joint concepts. In 2004, the COLMIL began work on the establishment of a Joint Special Operations Command (CCOPE) that would include the COESE with its commando battalion and *Lancero* group, joint urban antiterrorist units (AFEAU), an Air Force SF Group (ACOEA) with airborne snipers, and a 253-man, 3-company Navy SF marine infantry battalion.[80] In the summer of 2004, to the disappointment of American officials, the COLMIL ceased work on joint territorial commands. That October, Ambassador William B. Wood told Colombian officers that "the cooperation and coordination between US and Colombian forces could be better, and within the COLMIL, rivalries between forces jeopardize efforts to coordinate operations and share resources."[81] In December, Uribe intervened and directed the establishment of Joint Command No. 1—Caribbean (CCC).[82]

By the end of 2004, the security situation had continued to improve. JTF *Omega* had seized numerous FARC stockpiles, destroyed over 650 camps, and killed or captured almost 800 narcoterrorists.[83] The COLMIL and CNP strength continued to grow as units improved through continued training and combat experience. The Navy had doubled its riverine forces and established 2 Riverine Brigade—giving the Navy a marine infantry brigade and two riverine brigades.[84] The CNP continued to improve its presence in every municipality in Colombia. The AUC had begun to demobilize its estimated 15,800 armed members. For the first time, FARC strength had dropped significantly—from estimates of over 16,000 to 12,600 armed members. During the same time, the ELN remained a minor problem with about 3,500 members.[85] All the indexes of violence continued to drop and prospects for further improvements in 2005 seemed good. In fact, some talked of victory.

In the early months of 2005, the FARC responded with a series of attacks against isolated military and government targets—the greatest number since Uribe took office—in an effort to create casualties to undermine governmental efforts. In February, the FARC announced its "Plan Resistencia"—a strategy that ended its tactical withdrawal and initiated attacks against government targets.[86] To reduce its vulnerability, the FARC had reorganized into companies of 54 with squads of 12 personnel and focused on the use of snipers, landmines, and improvised explosive devices (IED) to create security force casualties.[87] In conjunction with increased attacks, the FARC sought to redirect its foreign affairs commission from lobbying abroad to working with nongovernmental agencies, to establish a nonmilitary presence in urban areas to infiltrate state organizations, to rebuild a military presence in areas cleared by the security forces, and to expand the political actions like the clandestine Bolivarian Movement for a New Colombia.[88] General Castellanos, the former JTF *Omega* commander who replaced Carreno as Army commander in November 2004, confronted the attacks that continued into 2006 and left more than 300 military dead in 2005 alone. According to a COLMIL analyst, the FARC in 2004 "were in a planned retreat. They were playing for time and waiting for the government to tire out. The military has not been able to neutralize the FARC, and now the rebels are trying to weaken Uribe and influence the next election." As with President Bush's "Mission Accomplished" banner, a Colombian Military analyst warned that the government had "prematurely declared victory."[89]

General Bantz J. Craddock, USSOUTHCOM commander since November 2004, provided Congress an overview of his area of operations in March. In Colombia, not only had crimes decreased in 2004—homicides by 16 percent—the lowest number since 1986, robberies by 25 percent, and kidnappings by 46 percent, terrorist attacks nationwide had dropped from 209 in 2003 to only 80, the lowest number since 1998. At the same time, military operations and demobilizations had reduced the strength of the narcoterrorist organizations. Craddock highlighted several other areas of improvement: support of civil affairs in reclaimed areas by Colombia's 13-ministry CCAI, increase in narcotics eradiation, improvement in counternarcotics interdiction programs, improvement in judicial cooperation, and increase in Colombian defense spending from 3.5 percent of gross domestic product (GDP) to 5 percent in 2004.[90] Military programs receiving US military assistance included the counterdrug BRCNA that was involved in both the US-led eradication program and the COLMIL-led Plan *Patriota*; a helicopter support program to train pilots, crew chiefs, and aviation mechanics; the infrastructure security program's 18 Brigade that had reduced pipeline attacks from 170 in 2001 to 17 in 2004; and the PATT Program that included 40 military personnel with plans to expand to 59 to assist selected Colombian regional joint commands, divisions, and brigades. Just over a third of DOD counterdrug funding requested for fiscal year 2006 was programmed for USSOUTHCOM and less than a third of that—$112 million—was for Colombia.[91] In 2004–2005, 7 SF Group deployed to Afghanistan and training support for USSOUTHCOM and Colombia declined. Generally, three ODAs provided training in Colombia—one with CCOPE units, one with CNP units—the *Junglas* and the *Carabineros*, and one with the Army Tactical Retraining Center

(CERTE)—supporting the 1-month retraining of Colombian Army combat units at three sites each supported by teams of 18 Colombian CERTE instructors.[92] US trainers introduced the after action review (AAR) process in 2005–2006, and the last CERTE rotations occurred in 2006.[93] In 2005–2006, the PATTs had been moved to work with the Army operations staff in Bogotá; with the 2, 3, 4, 6, and 7 Divisions; and with 18 Brigade and the Navy 2 Riverine Brigade.[94]

As with the adoption of other US military concepts, joint military commands remained a thorny issue within the COLMIL. Major General Montoya became the first commander of the Joint Carribean Command (CCC)—COLMIL's first regional joint command that consisted of his 1 Division, the Navy's riverine units in his area, and the Air Force's Combat Air Command No. 3. With minimal guidance and no transition plan, he assumed responsibility for air, land, and riverine operations in support of ongoing counterdrug programs and counter-narcoterrorist operations. As a functional joint command, CCC would remain a work-in-progress for years to come.[95] In April, Uribe ordered the creation of five joint territorial commands—Caribbean in the north along the border with Panama, Omega in the southeast along the border with Peru and Brazil, Central for the Andean region, Pacific for the southwest along the border with Ecuador, and Eastern along the border with Venezuela.[96] When four senior general officers raised questions to the Colombian Army commander about the utility of US-style joint commands and the wisdom of such a radical change in the midst of executing Plan *Patriota*, they were immediately retired.[97] In the controversy that followed, retired COLMIL officers voiced the concerns of many military officers. One of the four generals retired raised institutional issues: "no analysis of . . . political, constitutional, and military significance" of joint commands, lack of operational command "left [service chiefs] only with the responsibility of supplying troops and recruiting men," and "institutional fragmentation" similar to when a naval officer—"What can he do to lead . . . men from the army?"—had served as the second commander of the US-sponsored counterdrug JTF-S. A former COLMIL commander added the national issue: "Joint commands are a bad imitation of the American experience and the Colombian conflicts bears no comparison with that of the United States. . . . We cannot allow ourselves to apply an American organizational chart." General Tapias, the COLMIL commander under Pastrana, acknowledged, "Joint commands are necessary," but "the trouble lies in the strategy for implementing them. Quite often, what is beneficial for other countries is not beneficial for Colombia, because the conflict we have is unique to us." He added, "Let us hope that . . . [this decision is] backed by a thorough study of the threats from terrorism in the country because this is really a drastic venture."[98] Although the requirement for additional joint regional commands remained in place, during the remainder of Uribe's first term and into his second, the COLMIL joint commands remained JTF *Omega*, CCOPE, and CCC. In July, the Army established 7 Division and divided 1 Division area of responsibility in the north between these divisions and both became part of the CCC. (See Figure 9, The COLMIL areas of operations.) After mid-2005, the COLMIL oversaw the operations of CCC in the north, of JTF *Omega* in the southeast, and of CCOPE against national HVTs. The Army directed operations in 2, 3, 4, 5, and 6 Division areas and provided forces for COLMIL joint commands. The Navy had responsibility for the Caribbean Naval Forces Command, the Pacific Naval Forces Command, and those riverine forces not provided to CCC and JTF *Omega*. The Air Force provided air support to each of the joint commands and to the divisions.

After a couple of years of working together under expanded authorities, the Colombian–US relationship had evolved. It took years to overcome many of the misunderstandings, frustrations, and mistrust that had arisen in the pre-expanded authorities' period. As one senior Colombian official put it, "Arrogance ran both ways." Proud, under-resourced Colombian security force professionals struggling in a harsh internal war under Colombian conditions and constraints clashed with mission-oriented, we-are-here-to-help-you-in-our-own-way Americans. Both thought they knew best. Both tended to acknowledge the importance of human rights, strategic planning, MOD reform, jointness, intelligence-sharing, special operations, and procurement of critical assets. However, each tended to disagree on the form and content of these programs, on evaluation criteria, and on the time needed to produce results. American assistance helped, but to many it "distorted" the security force effort—reinforcing security force rivalries and dispersing critical assists such as helicopters—by pushing things not always needed, by insisting on things not always wanted, and by not always supporting things considered critical by the Colombians.[99] For example, for years the COLMIL had emphasized human rights training. Violations had dropped as the security forces increased in size and number of combat operations conducted. Yet, in spring 2005, the MILGP again had to encourage the Army to vet additional units. Units vetted by the DOS to US standards, not a drop in human rights violations, tended to be the US metric, and, for institutional and professional reasons, the Colombian Army had continued to resent and resist.[100] Afterward, Army resistance subsided and additional units were vetted. Over the years, many in the MOD concluded that the United States would not sell Colombia what it wanted.

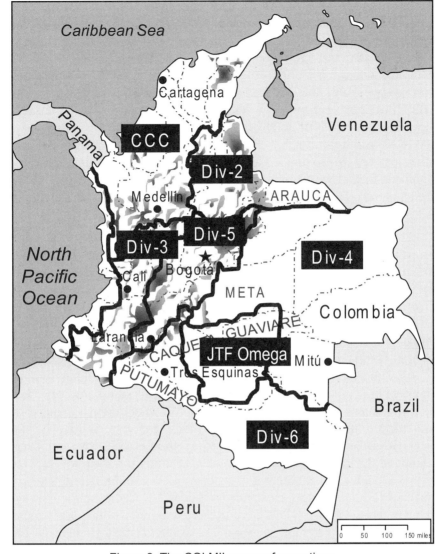

Figure 9. The COLMIL areas of operations.

Particularly prior to the expanded authorities, they believed the United States tended to push obsolete equipment—airplanes, helicopters, vehicles, and personal equipment considered good enough for Colombians—and failed to deliver items as programmed—"Plan Colombia" and ISS support packages. Considering the Foreign Military Sales (FMS) Program bureaucratic and unresponsive, the MOD procurement officials preferred, when permitted, to deal directly with the manufacturer to purchase critical items. Initially, American officials failed to see the importance of the *soldados campesino* program for providing local security. They encouraged the Colombian Air Force to acquire C-130 transport aircraft when it wanted to acquire Brazilian Super Tucano close air support aircraft, which it finally ordered in late 2005. Interestingly, while Colombian officials acknowledged the important role of US military assistance in sustaining combat operations, they stressed the critical importance of the interaction with US military personnel in developing and improving programs.[101] A former vice-minister of defense (VMOD) described the US military support as a "very light touch," one "never . . . involved in operational decisions." He concluded that US assistance "brings a rigor of training, focus, organization that we Colombians lacked. It has sped up what may have been a much slower process. US assistance brings decisiveness and helps us to make decisions."[102]

The growing security forces—COLMIL and CNP—continued to improve through individual and unit training, retraining, and combat operations. The professionals serving in the Colombian-developed mobile brigades and counterguerrilla battalions remained the fighting core of the Colombian Army. However, Army infantry battalions had grown in combat potential. The infantry battalion had increased to seven infantry and a support company. (See Figure 10, Colombian Army infantry battalion.) Six infantry companies had four platoons each. Each platoon was authorized 3 NCOs and 36 soldiers. Three companies had regular soldiers without high school degrees and three companies had professional soldiers. These six companies of trained combat soldiers, 24 platoons, had a strength equivalent to three counterguerrilla battalions. The seventh infantry company had *soldados campesino* platoons assigned based on the number of municipalities in the battalion area of operations that had one or more of these locally raised platoons. For example, a battalion commander in northern Colombia had nine municipalities in his sector—three with one platoon, three with two platoons, and three with no platoons. His seventh infantry company had nine platoons. The support company had an 81-mm mortar platoon, a security platoon, and a support platoon.[103] However, the expansion in manpower had not included larger or additional classes at the 4-year military academy or the 1.5-year NCO school. A severe shortfall of leaders, officers and NCOs, resulted with almost half of the Army units missing officers and with the Navy's riverine and naval infantry having a 1,200-officer shortage. Instead of a major, often a captain or senior lieutenant commanded counterguerrilla battalions. Sergeants or corporals ran most platoons. "Ten year soldiers are the ones making things happen," observed a PATT member, "but they can't be promoted."[104] Under the personnel system, NCOs could not become officers and professional soldiers could not become NCOs. Despite these problems, the military forces continued to improve their performance against the narcoterrorists and human rights violations did not increase proportionally—in fact they dropped—as combat operations had more than doubled.[105]

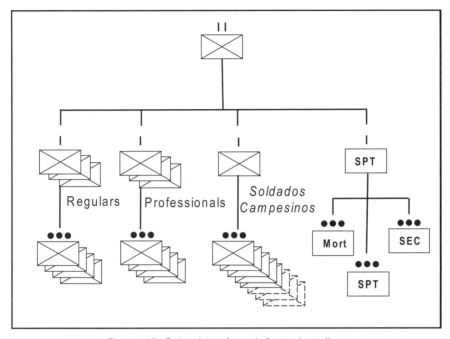

Figure 10. Colombian Army infantry battalion.

Plan *Patriota* not only included combat operations by JTF *Omega* and improved performance by jurisdictional military units, but after 2004 it had an increasingly capable HVT component. Early in his presidency, Uribe established a Colombian Joint Intelligence Council to consolidate analyses from his seven intelligence agencies—DAS, COLMIL, Army, Navy, Air Force, CNP, and Treasury.[106] Although far from perfect, it provided Uribe a means to focus agencies on specific HVTs. The creation of the CCOPE provided a national-level special operations force to attack HVTs. Through trial and error, both the intelligence—technical and human—and the capability to conduct intelligence-driven HVT operations improved. However, success against HVTs remained difficult to achieve. In September, after months of preparation to understand how and where a particular FARC front involved in narcotrafficking operated and training for a direct action force, local Army units established a grid around the targeted FARC unit. When the narcoterrorists tried to cover the escape of their leader, a pre-positioned CCOPE force captured him and five men. After talking with the captives, they released three to return to their comrades to encourage surrender. Leaderless, surrounded, and provided an option to surrender, every member of the FARC unit gave up without a fight.[107] The success of this operation proved an exception. Few operations made contact with their targets and fewer were as successful, but the pursuit of HVTs continued as intelligence, procedures, and capabilities improved.

At the end of 2005, the US counternarcotics "Plan Colombia" ended. After 6 years; $4.9 billion spent primarily on counterdrug programs that included 72 helicopters, a counterdrug brigade, the Air Bridge Denial Program, infrastructure security, the riverine program, CNP units, and aerial eradication; $1.3 billion spent on nonmilitary programs; and record eradication statistics with improved security, coca cultivation in Colombia had increased 15 percent since 2000 when the goal had been a 50-percent reduction.[108] Nevertheless, the US Congress agreed to support a three-stage Plan Colombia Consolidation Phase (PCCP). Under this program, no change in the level of US assistance would occur through fiscal year 2008. During the Uribe period, US assistance to the military and the police—both counterdrug and counterterror programs—had comprised 10 to 11 percent of the MOD budget. After fiscal year 2008, the programs would be consolidated, US funding reduced, and transferred to Colombia. In the last stage, nationalization or Colombianization, Colombia had ownership of the programs.[109] Despite the day-to-day problems and setbacks that sometimes obscured progress, the trends in Colombia continued to be positive. Security and local governance improved. Crime statistics continued to fall. From 2002 to 2005, homicides dropped 37 percent, kidnappings 72 percent, kidnappings with extortion 76 percent, and massacre victims 63 percent; terrorist acts declined from 1,645 to 611; and road mobility improved 45 percent.[110] The counterdrug programs—eradication, interdictions, and extraditions—all reached new highs. Transition continued to the new judicial system. Attempts throughout the year by the FARC to discredit Uribe and turn back security force gains failed. Uribe, as popular as ever, received legal approval to run for an unprecedented second term. The passage in July of a Peace and Justice Law facilitated the demobilization of the remainder of the paramilitaries. By the end of the year, the AUC demobilized 13,000 of an estimated 20,000 members. In mid-December, the ELN, reduced in strength to about 2,000, began talks in Cuba. FARC strength remained about 12,000.[111] If Uribe won the election in May, he would have another 4 years to build on the gains brought about by the

120

DSDP. Instead of being a year of impending change and new policies, 2006 could become one of continuity.

President Uribe provided unprecedented support to his security forces, but he demanded results and held commanders accountable for their organizations. Almost a dozen Army generals had lost their jobs by 2006 for reasons ranging from misusing funds to inadequate results against the narcoterrorists to disagreements over joint commands. Dismissals came suddenly and often publicly. In February, President Uribe dismissed General Castellanos—his Army commander and the highly-respected and successful commander of Operation Liberty I around Bogotá in 2003 and the first commander of JTF *Omega* from 2004 to 2005—after a serious hazing incident at an Army training facility. At the ceremony for General Mario Montoya, Castellanos' replacement as Army commander, President Uribe informed his audience, "It's difficult to defend the idea that we respect human rights . . . when we are accused of violating human rights inside the Armed Forces."[112] Dismissals for corruption and human rights reasons also occurred within the CNP. Perhaps the inability or unwillingness of security force senior leadership to fix responsibility and take prompt disciplinary action encouraged such Presidential action. Between dismissals and resignations during his first term as President, Uribe had three Ministers of Defense, two COLMIL commanders, and four Army commanders. Although many in the security forces questioned the wisdom and timing of some of these dismissals, few doubted the right of the President to take such action.

In March, General Craddock informed Congress that the "top priority" of USSOUTHCOM in Colombia had become the "safe return of the three American hostages" held by the FARC since early 2003. He praised the progress of the DSDP in strengthening Colombian democratic institutions and security—homicides had reached an 18-year low and attacks on towns had dropped 84 percent. Plan *Patriota* had disrupted the FARC support infrastructure, isolated its units, reduced its overall strength, and removed a score of mid-level commanders. In 2005, FARC losses totaled 400 killed and 445 captured. Ongoing demobilization and reintegration of narcoterrorists—particularly AUC organizations—had included over 16,000 *autodefensas* in 2005 and over 8,000 individual defectors from all 3 illegal armed groups—FARC, ELN, and AUC—since Uribe took office. Craddock highlighted the $4.5 billion MOD budget for 2006 and the interagency efforts of the CCAI. He stressed the continued importance of the expanded authorities to support Colombia's "unified campaign" against narcotrafficking and terrorists and of the personnel cap of 800 military and 600 civilians. He noted that the highest number of military personnel in country at any one time in support of "Plan Colombia" had been about 520.[113] At the end of September 2005, there had been 359 DOD personnel and 365 civilian contractors in Colombia.[114]

The May Presidential election dominated events in the first 5 months of 2006. The FARC tried but failed to disrupt the elections. The security forces provided protection for the population and continued to push FARC elements into the remote jungles along the borders of Ecuador, Venezuela, and Panama. Over a 3-year period, the largest source of violence and human rights violations—the AUC—completed its collective demobilization on 18 April. During that time, 30,150 *autodefensa* members—a higher total than government estimates—surrendered over 17,000 weapons, 117 vehicles, and 3 helicopters along with urban and rural properties. Under the Justice and Peace Law—a controversial law that did not provide amnesty but did provide for reduced sentences—over 600 faced accountability for their crimes before a Justice and Peace Tribunal. In addition to these procedures, Colombia extradited some of the notorious AUC drug traffickers to the United States for prosecution.[115] After 18 April, non-demobilized *autodefensas* became classified as criminals subject to apprehension and trial. In August, after further negotiations, a final AUC unit disbanded bringing the total demobilized in 38 actions to over 31,670 with over 18,000 weapons surrendered.[116] How well the reintegration program would work remained a concern. Although some former AUC-controlled areas came under FARC, ELN, or criminal influence, most Colombians saw AUC demobilization as another step toward better security and governance. Running as an independent candidate for an unprecedented second term as President, Uribe received 62 percent of the votes on 28 May for his second straight first-round victory in the least violent Colombian election in two decades.[117]

Democratic Security Consolidation Policy and Plan *Consolidacion*: 2006–2008

On 19 July, Uribe began establishing his second-term security team when he named Juan Manuel Santos as Minister of Defense. A former Minister of Finance, Minister of Foreign Trade, and naval officer, Santos became the first civilian Minister of Defense to have had military service in addition to extensive government experience. He promised to carry on with the DSDP by "correcting what needs correcting, strengthening what needs strengthening and continuing what needs continuing" and by making "life more difficult for . . . [the narcoterrorists] every day." Santos said his naval service allowed him to understand the Armed Forces and "demand more from them" since he knew "their foibles, their fears, their strengths, and their weaknesses." Disliking "making clean sweeps," he anticipated gradual changes in commanders.[118] Then, in mid-August, Uribe announced his security force appointments. The head of the Joint Chiefs of Staff, General Freddy Padilla—an engineer officer who had commanded 7 Brigade at Mitu, 5 Division at Bogotá in 2002, and served as Army inspector general—replaced Ospina as COLMIL commander. Admiral Guillermo Barrera assumed command of the Navy after Soto transformed it through his "Closing the Gap" strategy and the building of two riverine brigades during his 6-year tenure. General Jorge Ballesteros replaced Lesmes as Air Force commander. The recently appointed Army commander, Montoya, and the CNP commander, Major General Jorge Daniel Castro, retained their positions.[119] Describing Padilla as "a capable man in strategic command, an expert in intelligence . . . [who] has a sense of well-founded planning . . . [and] who could reconcile military know-how with strategic command," former Armed Forces commander General Tapias added, "Under his command the Armed Forces will be more united."[120]

In the fall, a series of incidents involving security force activities undercut public confidence and again tied military intelligence—as had been noted in 1997—to "secret behavior . . . linked to the violation of human rights . . . not under the control of any government office." Whether systemic or the result of pressure for positive results, poor coordination, or inadequate oversight, these concerns arose after a "friendly fire" incident between an Army unit and the police, faked terrorist attacks by Army anti-terrorism officers in Bogotá, "false positives" that presented civilian dead as dead narcoterrorists, leaked classified information to the media, and abuses in the payment of rewards to informants who provided unreliable information. In the resulting confusion, it became difficult to sort out the real from the false. In spite of efforts over the years to improve intelligence analyses and sharing, the problem remained that the lack of an intelligence system meant "the organization that has the information is the one that handles the problem . . . [which] leads to non-sharing of information." Minister Santos responded with a legislative proposal for intelligence sector reform, even though Colombia's Congress had failed to act on a similar proposal earlier.[121] After the dramatic security improvements during the first Uribe administration, the continued push for results created opportunities for abuse and a necessity for better supervision.

During this same period, increased narcoterrorist attacks added to the strains on the security forces. Through a series of attacks aimed at discrediting the security forces—bombings that included the Nueva Granada Military University in Bogotá and an assault against the isolated municipality of Tierradentro—the FARC reminded Colombians that it remained a threat. To attack Tierradentro, the FARC moved roughly 450 guerrillas from three fronts into an area formerly under AUC control. The less than 60-man police garrison suffered 17 killed, but prevented the FARC from occupying the town.[122] Immediately, the CNP reinforced the garrison and the Army moved forces that engaged the FARC elements. Putting these attacks into perspective, the COLMIL commander reminded the public of the security improvements over the previous 4 years— FARC strength dropped almost a third to roughly 11,000, terrorist attacks occurred less frequently with less result, completion of the "most successful demobilization process" in Colombian history, and a steep decline in crime statistics. Addressing the "false positives," he emphasized the care given to the orders issued and to the training provided military personnel to meet the goal that all "actions . . . correspond, day after day, to the guidelines of the Constitution and the Law" which, he noted, "is the only way to ensure that the Colombian people will continue to walk by our side." Padilla acknowledged two failures—not capturing a member of the FARC Secretariat and not rescuing the political hostages. "This does not mean . . . that we are not trying," he added because in 360 rescue operations over 480 captives had been released without a single hostage injury or death. Padilla's message noted that "every day we are performing in an increasingly more professional and technical manner and in keeping with the Constitution and the law" while the narcoterrorists, despite the recent attacks, were being worn down and dispersed.[123]

By the end of 2006, the Colombian security forces had grown in size, complexity, and capability. During Uribe's first 4 years (2002 to 2006), security force strength had grown 30 percent from 313,000 personnel in 2002—203,000 COLMIL and 110,000 CNP—to 406,000 (+30 percent)—276,000 COLMIL (+32 percent) and 139,000 CNP (+26 percent). The Army grew 35 percent from roughly 148,000—89,000 regulars without degrees, 59,000 professionals, and no *soldados campesinos*—to over 200,000 with 96,000 regulars (+8 percent), 79,000 professionals (+34 percent), and 25,000 *soldados campesinos*. As in the preceding years, in 2006 the Army continued to increase the number and types of units it needed to counter the narcoterrorist threat. (For an overview of some of the Colombian Army expansion, see Table 4, Growth of Colombian Army units, 2002–2006.) Among the units created, three provided new capabilities. To counter the FARC snipers, the Army created sniper detachments. To provide each division commander a Reserve, the Army established a counterguerrilla battalion-size division reaction force (FURED) commanded by a major. In November, the Army organized a 3-BRIM, light division-equivalent

Fuerza de Accion Decisiva or Decisive Action Force (FURAD)—in effect a second FUDRA—to provide the COLMIL a strategic reaction force.[124] In December—after a process than began in 2001 to replace 24 combat aircraft and a year after placing its order—the Air Force received its first 5 of 25 Brazilian-made, day-and-night capable EMB-314 Super Tucano close air support aircraft equipped with Israeli avionics. This aircraft provided greater loiter time, quicker responsiveness, and more accurate engagement of ground targets.[125] In an effort to improve performance, all HVT security force units came under a single office in the MOD. During operations in 2006, at least six senior FARC leaders, including a General Staff member, died.[126] With the demobilization of the AUC and almost 2,000 narco-terrorist desertions in 2006, the security forces could concentrate in 2007 on a FARC that had less than 12,000 members, on the ELN with a strength of about 2,000, and on criminals and drug-traffickers.[127]

Table 4. Growth of Colombian Army units, 2002–2006[128]

Colombian Army	2002	2003	2004	2005	2006	TOTAL
Divisions	1			1		2
Decisive Action Force (FUCAD)					1	1
Territorial Brigades	1	1	1	3		6
Mobile Brigades (BRIM)	2	1	3	3	3	12
Division Reaction Forces (FURED)					7	7
Urban Antiterrorism Special Forces Units (AFEUR)	10	1				11
Special Operations Command (COESE)		1				1
Lancero Group		1				1
Counterguerrilla Battalions (BCG)	10	4	8	12	14	48
Mountain Battalions	2	2	1		1	6
Intelligence/Counterintelligence Regions		4		1		5
Technical Intelligence Units		2				2
Plan *Meteoro* Companies	7				2	9
Instruction/Training Centers				20		20
Sniper Detachments					41	41

Building on the successes of the DSDP from 2002 to 2006 and supporting Uribe's National Development Plan 2006–2010, "Community State: Development for Everyone," Minister Santos published a new national defense policy, Democratic Security Consolidation Policy (DSCP), in early 2007. The DSCP moved beyond the focus on regaining territorial control by the security forces to "social recovery of that territory through integrated state action"—a whole-of-government effort to establish security, governance, rule of law, and social programs. The "virtuous circle of security" concept explained the relationship between security improvements and economic and social development. (See Figure 11, Virtuous circle of security.) A three-phased territorial consolidation strategy and an Integrated Action Doctrine—in which the security forces worked with the CCAI where it was present and with local civilian officials where it was not—provided the foundation of the DSCP. The DSCP implementing principles stated that the security force would work to establish security and peace, act within the law, maintain a continuous presence, provide security to permit the rule of law, maintain flexibility and adapt to the

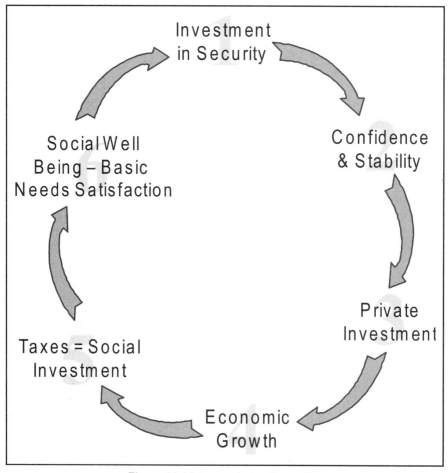

Figure 11. Virtuous circle of security.

security threats, and coordinate with other security forces—"Each one of the Armed Forces will intensify its process to adapt to the joint operations doctrine," and coordinate with other state agencies. The DSCP established five strategic objectives: consolidate territorial control and strengthen the rule of law throughout the national territory; protect the population and maintain the strategic initiative against all threats to citizen security; drastically raise the cost of trafficking in drugs; maintain modern, efficient, and legitimate security forces with earned public support and confidence; and sustain the downward trend in all urban crimes. To implement the DSCP, 28 plans, programs, and initiatives had been developed.[129] How the security force commanders embraced the nonsecurity aspects of the DSCP remained to be determined.

To implement the DSCP, the COLMIL replaced Plan *Patriota* with Plan *Consolidacion* or Consolidation—initially called Plan *Victoria* or Victory. The general military strategic concept—"execute simultaneous joint operations across the entire country"—included consolidating governance to support legal economies, social investment, and a functional justice system; working with social investment and development agencies; working jointly with the CNP on all security and counterdrug efforts; and continuing the transformation and strengthening of the COLMIL. The three-phased consolidation strategy—clear, hold, and state buildup—sought to align the security, social, and counternarcotics efforts. In the first phase, clear, "red" areas with active illegal armed groups underwent a military-intensive effort to drive out these armed groups and to establish territorial control. Hold, the second phase, focused on police and military efforts to sustain security in "yellow" areas to attract government agencies for the establishment of governance and social programs. In the third phase, state buildup, stabilized or "green" areas received intensive political and social efforts to consolidate state authority and to establish all state agencies and public services. The COLMIL mission stressed weakening the narcoterrorist will to fight, forcing narcoterrorist demobilization and disarmament, protecting the population, securing the national economic infrastructure, and consolidating the rule of law. Breaking the will of the narcoterrorists and reducing their offensive capacity implied a change in the metrics of success—from killed guerrillas to surrendered or captured. Four principles governed joint operations: unity of effort through the rational use of integrated and coordinated forces; synchronization from matching operations in time, space, and purpose; synergy from teamwork; and flexibility from an ability to adapt to the operational environment. According to the plan, military joint commands increased capability, decentralized employment, improved synergy through integration, improved employment and control, improved counterdrug interdiction operations, and reduced the offensive capacity of the narcoterrorists.[130] Just as it took time to build the forces necessary and experience to refine the operations that made Plan *Patriota* successful, Plan *Consolidacion* would require a multiyear, trial-and-error effort to approach its goals.

Confirming his predecessor's focus, Admiral James Stavridis, the USSOUTHCOM commander in October 2006, stated that freeing the American hostages was his "first and foremost" priority. In addition, the security situation in Colombia remained an area of focus and of great success—the result of Colombian interagency efforts and US resources and support—given that 10 years before "beheadings, kidnappings, torture, and bombings occurred essentially daily." "Since 2000," Stavridis noted, "Colombia stands out as a true reflection of what steady partnership with the US can achieve." He saw this reflected by sustained combat operations, the development of actionable intelligence, the protection of critical infrastructure, and improved civil-military cooperation—all performed within the "norms of international human rights and the rule of law." Stavridis emphasized the importance of training, particularly human rights training, that the Western Hemisphere Institute of Security Cooperation (WHINSEC) provided. He promised to "continue to bring innovative and experimental capabilities under development into Colombia"—unmanned systems, detecting objects under dense foliage, data fusion, biometrics— and "near real time support" from such "cutting edge technologies." He highlighted several programs—logistic support to JTF *Omega* and Plan *Consolidacion*, logistics support for helicopter aircrew and maintenance personnel, fielding of "Midnight Express" boats to the Navy for interdiction, and working with 2 Riverine Brigade. Stavridis asked for Congressional support for a proposed Center for Excellence in Human Rights noting that USSOUTHCOM was the only combatant command with a dedicated human rights program.[131] In 2006—after resisting US vetting procedures for years—the new MOD team, and the Colombian Army in particular, embraced human rights vetting. In time, instead of a short list of vetted units, the MILGP maintained a shorter list of nonvetted units.[132] The MILGP continued to provide logistics, technical, and training support for a variety of programs that addressed ground operations, helicopter operations, maritime interdiction, riverine operations, air support, intelligence, joint, and civil-military issues. In 2007, most PATTs—now part of the MILGP Field Liaison Groups—moved to support units in JTF *Omega* where they coordinated US support and provided information on ongoing operations.[133] Discussions began on the multiyear transition of US military assistance and counterdrug programs to Colombian ownership scheduled for after fiscal year 2008. Ambassador William R. Brownfield arrived in August to oversee that process.

The past continued to influence the present. As governor of Antioquia, President Uribe had established CONVIVIR self-defense units to address the security situation in his department. Allegations of ties to the *auto-defensas* by Uribe and by some of his supporters, to include members of Congress, created a scandal that exposed a relationship between the para-militaries and elements in the political and economic elite. Confirmation of these charges against some resulted in convictions and prison. Security force personnel faced similar charges. The Army commander faced alle-gations that his brigade had worked with the *autodefensas* in reestablish-ing government control of a neighborhood in Medellín known as Comuna 13 during Operation Orion in October 2002.[134] During the annual DOS human rights certification process, the Colombian Attorney General's Office reported that since 1 January 2002 it had initiated disciplinary pro-ceedings against 54 military personnel for presenting 94 dead civilians, killed accidentally or intentionally, as dead guerrillas. Half of these cases remained under investigation, but half had been recommended for disci-plinary action. In addition, six cases had been filed against Army person-nel in five different departments alleging their "participation in the deaths of individuals whom they later presented to the public as guerrillas and on whom they reportedly planted firearms to confirm the accusation that they were subversives."[135] Given the importance of results, the shortage of officers and NCOs, the increase in combat operations, and the belief that successful combat operations meant dead guerrillas, the fact that these "false positives" occurred should have been no surprise.

Despite the problems addressed above, security force operations continued against the illegal armed groups throughout the country. JTF *Omega* focused on clearing and holding the municipalities in its area of operations in the heart of what had been a FARC-dominated region. With improvements in targeting (human intelligence, technical intelligence, and intelligence sharing) and in attack capabilities (CCOPE, Air Force Super Tucanos, helicopter-borne Army units), successes against leader-ship targets complemented the successes of local operations. By the end of the summer, a senior FARC commander—El Negro Acacio—had been killed, a major drug trafficker—Don Diego—had been captured, and the camp of another FARC leader destroyed. "Nowadays there is no safe place for the FARC leaders," noted the COLMIL commander because it had "acquired technological capacity that they did not have previously, and their organization and equipment, with the support of friendly countries, is giving fruit. Day or night, there is no place where the Armed Forces cannot venture" without surprise and decisiveness.[136] Under this pressure of sustained and increasingly successful combat operations by the security forces, FARC units diminished in strength, became isolated from one another, and moved to the border areas for safety. In June, a FARC front holding 12 legislators from the Valle Del Cauca department hostage since their capture in Cali in 2002 killed 11 of them. To justify the action, the FARC blamed a botched hostage rescue when no operation had occurred. It appeared that the FARC commander either feared such an attempt or no longer possessed the means or the will to hold these important hostages.[137] The security force operations continued to stress a weakened FARC organization and to disrupt narcotrafficking activities.

Plan *Consolidacion* placed greater focus on the integrated action of the security forces and civilian government organizations. Before 2007, the CCAI had attempted to coordinate pilot programs in 11 priority regions that encompassed 58 municipalities. Some of the programs worked well, others failed to produce useful results. The CCAI's number one priority region— the *Zona Sur* or south zone included JTF *Omega's* area of operations— had proved a difficult area in which to make progress.[138] Under the DSCP, Santos explained, "We have to go to the more remote areas, where we have drug trafficking and illegal groups and we have poverty" and the military had to work with civilian agencies to provide healthcare, schools, and infrastructure improvements—something that had not happened in that area before.[139] In August, the MOD sponsored the CCAI *Plan de Consolidacion Integral de la Macarena* (PCIM), a project for six municipalities in Meta department. In addition to security and drug eradication, the PCIM focused on five areas: governability, institutional development, and citizenship; territorial order and property rights; infrastructure and connectivity; access to public goods and social services; and business and economic development. Although JTF *Omega* had operated in this area since 2004, progress against 27 and 43 FARC Fronts—among the largest and best armed—improved in 2007 after the focus shifted from clearing or killing guerrillas to holding the areas cleared. During the next year in the PCIM areas, FARC morale dropped, desertions rose, and its drug-based economic support declined.[140] As with many security and development programs in Colombia, sustainability and the role of the security forces in nonsecurity functions remained a question.

By the end of 2007, the narcoterrorist threat had continued to decline—FARC strength had dropped to 9,500 and a 2,000-member ELN still talked negotiations. However, new illegal groups with ties to narcotrafficking arose in areas. The local department and municipality elections in October proceeded without interference with only 25 candidates killed nationwide.[141] Crime statistics dropped and narcoterrorist desertions increased. During the year, narcoterrorist units had not attacked a town.[142] The MOD reported 67 percent of the population in "green" or state building areas, 19 percent in "yellow" or hold areas, and only 14 percent in "red" or clear areas.[143] Counternarcotics operations increased in number and in results. Manual eradication—now the preferred Colombian eradication method—increased 50 percent. Extraditions to the United States reached a record 164. To create additional units to support holding operations, the CNP reorganized its 150-man *Carabinero* squadrons into 120-man units.[144] Despite these improvements and operational successes by the security forces, the FARC remained undefeated.

Nevertheless, an undefeated FARC did not mean a capable FARC. Not only had FARC strength declined to almost half its peak strength—the result of increased deaths and desertions that included mid-level leaders—but at least 20 fronts, to include the elite Teofilo Forero mobile column, had become ineffective or ceased to exist. Among these units, the Teofilo Forero column had been reduced to 54 of 200 members, 52 Front near Bogotá disbanded when its last three members moved to other fronts, the 200-man 40 Front in Meta disbanded, 26 and 31 Fronts had fewer than 50 of 220 members, and 19 and 35 Fronts in the north disappeared from deaths and desertions. Security force operations forced most fronts to disperse into small groups. Improved sharing of real-time, actionable intelligence between the military services and the increased bombing accuracy of the Super Tucano aircraft permitted the operations that killed two important FARC commanders—El Negro Acacio in September and Martin Caballero in October. With its advanced firing systems and accurate bombing, the Super Tucano permitted ground forces to arrive at the target almost immediately after engagement. A Colombian Army general said, "The use of what are known as intelligent weapons improve the effectiveness of the operations and eliminated the use of wide area bombing that made more noise than caused deaths."[145] Capabilities had improved to the point that if a HVT could be located by human or signals intelligence, it could be targeted. If it could be targeted, it could be attacked promptly and accurately from the air. Ground forces could be pre-positioned so that immediately after the air attack, these units exploited the site to assess the damage and to gather intelligence about other potential targets. To counter this threat, FARC units dispersed in small groups and minimized their use of electronic communications equipment, which further hampered its command and control.

In January 2008, a Colombian analyst acknowledged that the military consolidation strategy—based on a larger budget, increased personnel, and the payment of rewards to informants—had "won a level of initiative . . . managed to get the guerrillas to retreat, and . . . managed to get the government to somewhat regain territorial control" while seeking to demobilize the illegal armed groups, capture their members, and maintain a presence in every municipality. Despite some success against mid-level leaders, he noted the security force failure to strike members of its principal HVT—the FARC Secretariat.[146] That changed with Operation *Fenix* or Phoenix on 1 March. In a carefully planned, professionally executed joint operation involving CNP, Army, and Air Force units, the COLMIL struck the camp of Raul Reyes—the FARC number two and chief spokesman—just inside the border of Ecuador. After a bombing attack, ground forces assaulted the site and found the body of Reyes and a laptop computer with valuable intelligence about the FARC and its supporters. Military analyst Alfredo Rangel commented:

> His death comes after a lot of defeats at the hand of the army, the capture of leaders, the killings of leaders, the loss of territory, the reduction in finances, the reduction in military capability. So this could be very significant, in the perception of the guerrillas may have about the possibility of military success in the future. This shows that its military project has no possibility of triumphing in Colombia.[147]

Within days, a second Secretariat member and the principal FARC peace negotiator, Ivan Rios, died in northwest Colombia at the hand of his bodyguard who presented Rios' right hand—along with a laptop computer—as proof of death when he claimed a reward of over $2.5 million.[148] Although it would be 2 months before the news broke, the long-time FARC leader, Manuel Marulanda, died of a heart attack on 26 March making Alfonso Cano the FARC commander.[149] Within less than a month, three of the seven-member Secretariat had died from military attack, treachery, or natural causes. Despite these successes, the FARC was able to fill the three vacancies on the Secretariat.

Citing the progress in security, judicial matters, and political transparency, in March Admiral Stavridis described Colombia as "on the brink of winning its peace and making its successful gains against terrorism and social disorder irreversible." During the 3-year transition to Colombian-owned programs, US support of the Armed Forces and the interagency efforts in former FARC-controlled regions would focus on training, mobility, and sustaining "key infrastructure programs" that provided "long-term self-sufficiency" supported by a 12-percent increase in the Colombian defense budget that would provide $3.7 billion between 2007 and 2010.[150] During the Uribe period, over 55 percent of the US funding—CNP and COLMIL counterdrug—continued to support counterdrug programs. The remaining 45 percent was divided almost evenly between nonmilitary and Army programs. (See Table 5, US assistance to Colombia, 2003–2008.) The majority of the CNP assistance supported the DIRAN's counterdrug operations, its elite *Junglas*, and the *Carabineros*—about 15,000 personnel or 10 percent of the national police. The majority of the COLMIL training and assistance focused on counterdrug units—BRCNA, aviation, and riverine; on infrastructure security; and on elite units like the CCOPE and JTF *Omega*—likewise a small percentage of the COLMIL manpower.[151] The majority of the funding—military and nonmilitary—came from the Colombian Government. To support the transfer of programs to Colombia and to make the security gains irreversible, the MILGP identified 20 programs grouped into 8 program areas that it listed by funding priority: rotary-wing operations, ground operations, riverine operations, fixed-wing operations, joint initiatives, naval interdiction, governability, and intelligence and communications.[152] A Government Accounting Office (GAO) report on the US transition efforts concluded—as with earlier programs in Colombia—that efforts were "not guided by an integrated plan that fully addresses the complex mix of agency programs, differing agency goals, and varying timetables for nationalization."[153]

In March, the issue of "false positives" arose again as a debate surfaced within the MOD between generals who supported aggressive operations that focused on killing narcoterrorists or guerrillas and those who favored metrics other than body counts. The chronic but isolated problem of military personnel killing civilians—extrajudicial killings—and passing them off as guerrilla dead had grown as the pressure for results clashed with improved security that made results more difficult. Most victims had been peasant farmers, but the number of kidnapped urban poor killed had grown. Over 650 cases between mid-2003 and mid-2007 that involved over 1,000 dead were under investigation. Minister Santos acknowledged the problem: "I've told all my soldiers and policemen that I prefer a demobilized guerrilla, or a captured guerrilla, to a dead guerrilla." He added, "I have said this very clearly: the soldier who commits a crime becomes a criminal, and he will be treated as a criminal." In addition, Plan *Consolidacion* had emphasized the demobilization or surrender of narcoterrorists. This implied a metric of success that gave priority first to surrendered, second to captured, and last to killed guerrillas. However, many in the military supported aggressive, decentralized operations and shared the view espoused by the Army commander: "What's the result

Table 5. US assistance to Colombia, 2003–2008[154]

Dollars in millions	FY 03	FY 04	FY 05	FY 06	FY 07	FY 08 (Est.)	TOTAL
Promote Social and Economic Justice	**$125.7**	**$126.4**	**$124.7**	**$130.4**	**$139.8**	**$194.4**	**$841.4**
Alternate Development	$60.2	$59.8	$70.7	$72	$68.2	$119.7	$450.6
Internally Displaced Persons	$41.5	$42.6	$32	$30.7	$31.1	$35.3	$213.2
Demobilization/Reintegration				$8.9	$15.7	$18.3	$42.9
Democracy and Human Rights	$24	$24	$22	$18.8	$24.8	$21.1	$134.7
Promote Rule of Law-Judicial Reform and Capacity Building	**$27**	**$9**	**$7.3**	**$10.5**	**$7.8**	**$39.4**	**$101**
CNP	**$164.5**	**$172.2**	**$190.9**	**$204.5**	**$217.7**	**$155**	**$1104.8**
Eradication	$63.7	$44.2	$82.5	$81.7	$82	$66.5	$420.6
Air Service	$62.3	$71.2	$70	$70.5	$69	$52.5	$395.5
Interdiction	$21	$41	$16.9	$16.5	$16.5	$16.5	$128.4
Police Presence—Conflict Zones	$15.5	$13.8	$20.1	$19.4	$18.7		$87.5
Other	$2	$2	$1.4	$16.4	$31.5	$19.5	$72.8
COLMIL Counterdrug	**$203.3**	**$268.1**	**$249.9**	**$213.4**	**$222.4**	**$182.2**	**$1339.3**
Air Interdiction	$8	$7.1		$4.6	$18.8	$10	$48.5
Coastal/River Interdiction		$26.2	$11.8	$19.1	$19.2	$13	$89.3
Counterdrug Funding	$195.3	$234.8	$238.1	$189.7	$184.4	$159.2	$1201.5
Colombian Army	**$240.1**	**$177.3**	**$144.9**	**$169.4**	**$151.3**	**$86.1**	**$969.1**
Aviation	$140.8	$155.2	$127.5	$143.2	$129.6	$69.7	$766
Ground Forces	$6.3	$18.1	$13.4	$22.2	$17.7	$16.4	$94.1
Infrastructure Security	$93	$4	$4	$4	$4		$109
TOTAL	**$760.6**	**$753**	**$717.7**	**$728.2**	**$739**	**$657.1**	**$4355.6**

of offensives? Combat. And if there's combat, there are dead in combat." Successful combat meant dead guerrillas. In fact, soldiers received extra pay or time off for kills in combat, but for those captured or who surrendered no such rewards accrued—just the burden of securing and feeding them until they could be transferred to the police. A former soldier said, "The army gives prizes for kills, not for control of territory."[155] Although the COLMIL plan emphasized demobilization of the narcoterrorists, many Army operations continued to stress its core function and its metric of success for the past 44 years—dead guerrillas.

After the military successes against Secretariat members in March, Alfonso Cano struggled to gain control of the FARC. First, the constant pressure of security force operations wore the FARC down. A FARC commander wrote that the Army "doesn't let up. The number of troops is enormous. Sometimes we eat once a day." This stress led to despair, increased desertions, and fear of betrayal. Second, security force capabilities and successes disrupted FARC command and control. Afraid of detection from the use of mobile phones, the FARC reverted to foot messengers. In May, Karina—a one-eyed, 24-year veteran and Front female commander—surrendered. She announced that she had not communicated with her commanders for over 2 years and added, "Everywhere we went, the army was there. We couldn't sleep in one place for more than one night." Third, the government used some of those who surrendered as guides to locate their former units and used others to make radio and television appeals to their comrades. A FARC commander complained that defectors "are constantly on the air" and that in his unit 10 members deserted and 4 of them surrendered to government forces. Fourth, Mono Jojoy—the FARC military and the Southern bloc commander—had became a rival for FARC leadership. And, as if he had no other problems, the improving security force intelligence net—using a nationwide network of civilian informants, aerial intelligence systems, and improved intelligence sharing between the Armed Forces and the police—increased the vulnerability of all HVTs.[156]

On 2 July the Colombian Army launched one of the boldest, most masterful hostage rescue operations ever conducted—Operation *Jaque* or Check. After months of tracking a group of FARC political hostages that included former Presidential candidate Ingrid Betancourt captured in 2002, three American contractors—Keith Stansell, Thomas Howes, and Marc Gonsalves—taken in 2003, and 11 long-held security force personnel, Colombian intelligence personnel proposed an unusual scheme using informants and simulated communications between senior FARC leaders and the 1 Front commander—Cesar—who held the prisoners. Convinced that Cano wanted to move the hostages, Cesar brought them together, met a civilian helicopter that included aid workers, press personnel, and a FARC member he knew, loaded the hostages on the helicopter, and agreed to fly to the rendezvous. He had no idea that the white MI-17 was an Army helicopter or that the unarmed civilians onboard were security force personnel executing a carefully planned, thoroughly rehearsed rescue. Without a shot fired, Operation *Jaque* dealt the FARC another crippling blow—loss of its political hostages and capture of its 1 Front commander.[157] Taking pride in the success gained and personally aware of the risks run, a senior Colombian officer remarked, "People will always say it's impossible that Colombians did this on their own. Of course the gringos have to be present."[158] Ambassador Brownfield acknowledged, "This mission was a Colombian concept, a Colombian plan, a Colombian training operation, then a Colombian operation." Then he added the United States "had been working with them more than five years on every single element that came to pass that pulled off this operation, as well as the small bits that we did on this operation."[159] Few denied the critical support role played by American assistance over the years or the intense US military focus on rescuing the American hostages, but few militaries would have considered, much less executed, this Colombian operation.

Changing an institutional culture is difficult; it takes time, often decades, to go from saying something to having that something accepted to having that something done routinely within a military organization. The process may begin with training or directions from the top, but it takes much more—things like constant emphasis, widespread supervision, strict accountability, rewards for success, and punishments for failure—to change the way a military thinks about its roles, and thus what it does. On 3 October, the MOD convened a special military commission to investigate the deaths of 11 urban poor who were killed as guerrillas in northeast Colombia. Before the end of the month, President Uribe fired 20 Army officers—3 generals that included 2 division commanders, 4 colonels, 7 lieutenant colonels, and 6 others—along with 7 soldiers. In the largest human-rights related dismissals during his presidency, Uribe said, "In some instance, there has been negligence in the army, and that has permitted some people to involve themselves in crimes, which in some regions end in the killings of innocents to show success against the criminals." Santos declared that the 27 were "administratively responsible" through acts of commission or omission and repeated for his commanders, "The measurement of success cannot be body count. Our men have to be commended for controlling territory, for the number of demobilizations that are registered. I have said it many times. I prefer a demobilized guerrilla to a combat kill."[160] Not consulted about the firings, the Colombian Army commander resigned to be replaced by General Oscar Gonzalez.[161] By 2008, most observers agreed that the security forces had made tremendous progress in the human rights area. A Colombian analyst described their "theoretical emphasis the best in the world, better than Switzerland," but acknowledged, as in many places, it was not always done well in practice.[162] What some saw as a chronic problem still uncontrolled, others viewed as a regrettable but isolated incident that received prompt, decisive action.

Improved performance against the narcoterrorist groups—and increased security—came from larger security forces, better training, sustained combat operations, and the consolidation strategy. By the end of 2008, Colombia appeared to be on the brink of success in its struggle against these groups. The FARC numbered less than 9,000 members and the ELN remained at about 2,000.[163] Building on the Pastrana increases, under Uribe the security forces had grown from 313,000 in 2002—203,000 COLMIL and 110,000 CNP—to almost 432,000 in 2008—286,000 COLMIL and 146,000 CNP. The 212,000-member Army had roughly 101,000 regulars, 86,000 professionals, and 25,000 *soldados campesinos*.[164] In addition, the whole-of-government consolidation effort moved forward under Uribe's DSCP. Although the immediate threat of defeat no longer existed as it had in 1998, questions remained about the ability to further reduce crime, about the sustainability of the consolidation programs, about the MOD reform effort and joint commands, about the roles and missions of the security forces, about the ability of the military to do integrated action or civil-military projects, and about the road Colombia might take after President Uribe. Many of these questions may be answered by the end of the second Uribe presidency in 2010 and after the transition of US-funded programs. Then one can better evaluate if the security gains and increased governance can be considered irreversible.

Notes

1. Republic of Colombia, *Democratic Security and Defense Policy*, Republic of Colombia, 16 June 2003, 7.

2. "Colombian Minister Stresses Importance of Role of State in Security Issues," *BBC Monitoring Americas—Political*, 21 October 2002, http://proquest. umi.com/pqdweb?index=19&did=217198381&SrchMode=2&sid=15&Fmt= 3&VInst=PROD&VType=PQD&RQT=309&VName=PQD&TS=1236966153& clientId=5904 (accessed 4 November 2008).

3. James T. Hill, "Statement before the House Armed Services Committee on the State of Special Operations Forces," 12 March 2003, http://armedservices. house.gov/comdocs/openingstatementsandpressreleases/108thcongress/03-03-12hill.html (accessed 18 March 2009).

4. Nina M. Serafino, "Colombia: The Uribe Administration and Congressional Concerns," Congressional Research Service Report, 14 June 2002. Serafino concluded that from a US Congressional perspective, Uribe's three major challenges were (1) counternarcotics, (2) economic, and (3) security. Uribe probably would have agreed with security and economic; Yadira Ferrer, "Colombia: Uribe Launches Controversial Network of Informers," *Global Information Network*, 9 August 2002, http://proquest.umi.com/pqdweb?did=32 1047661&sid=1&Fmt=3&clientId=5094&RQT=309&VName=PQD (accessed 4 November 2008).

5. Myles R.R. Frechette, *Colombia and the United States—The Partnership: But What is the Endgame?* (Carlisle Barracks, PA: Strategic Studies Institute, February 2007), 17.

6. "Colombian Minister Stresses Importance of Role of State in Security Issues."

7. Ferrer, "Colombia: Uribe Launches Controversial Network of Informers."

8. David Adams, "President Declares Emergency as 100 Die in Colombia," *Times*, 13 August 2002, http://proquest.umi.com/pqdweb?index=13&did=14874 8841&SrchMode=2&sid=1&Fmt=3&VInst=PROD&VType=PQD&RQT=309& VName=PQD&TS=1245765801&clientId=5904 (accessed 4 November 2008).

9. Juan Forero, "Burdened Colombians Back Tax to Fight Rebels," *New York Times*, 8 September 2002, http://proquest.umi.com/pqdweb?index=1&did= 163053271&SrchMode=2&sid=2&Fmt=3&VInst=PROD&VType=PQD&RQT =309&VName=PQD&TS=1245780557&clientId=5904 (accessed 4 November 2008).

10. Rachel Van Dongen, "'Zones' Suspend Colombian Rights; Military Put in Charge of Security, Given Authority Over Civilians," *Washington Times*, 3 December 2002, http://proquest.umi.com/pqdweb?index=3&did=250009071 &SrchMode=2&sid=3&Fmt=3&VInst=PROD&VType=PQD&RQT=309&V Name=PQD&TS=1245781832&clientId=5904 (accessed 4 November 2008).

11. United Nations, "Report of the United Nations High Commissioner for Human Rights on the Human Rights Situation in Colombia," United Nations

Commission on Human Rights, 24 February 2003, 43, http://www.unhchr.ch/Huridocda/Huridoca.nsf/0/1304674285b7eb3bc1256cf5003906fb?Opendocument (accessed 19 March 2009).

12. "Defense Minister Says Security Forces Must Be Present Throughout Colombia," *BBC Monitoring Americas—Political*, 26 August 2002, http://proquest.umi.com/pqdweb?index=0&did=155320491&SrchMode=2&sid=1&Fmt=3&VInst=PROD&VType=PQD&RQT=309&VName=PQD&TS=1245775397&clientId=5904 (accessed 4 November 2008).

13. "Colombia: General Reacts to Public Reprimand of Armed Forces by President Uribe," *BBC Monitoring Americas—Political*, 29 August 2002, http://proquest.umi.com/pqdweb?index=0&did=156290021&SrchMode=2&sid=1&Fmt=3&VInst=PROD&VType=PQD&RQT=309&VName=PQD&TS=1245779069&clientId=5904 (accessed 4 November 2008).

14. Darren D. Sprunk, "Transformation in the Developing World: An Analysis of Colombia's Security Transformation" (Monterey, CA: Naval Postgraduate School Thesis, September 2004), 46 and 57.

15. US Department of State, "International Narcotics Control Strategy Report for 2002" (Washington, DC: US DOS, 1 March 2003), http://www.state.gov/p/inl/rls/nrcrpt/2002/index.htm (accessed 1 October 2008).

16. "Colombian Minister Stresses Importance of Role of State in Security Issues."

17. Sprunk, "Transformation in the Developing World," 72.

18. Scott Wilson, "Colombia Poised to Install Leader as Rebels Attack; Dozens Dead in Wide-Ranging Offensive on Eve of Hard-Liner's Inauguration," *Washington Post*, 7 August 2002, http://proquest.umi.com/pqdweb?index=3&did=146818411&SrchMode=2&sid=1&Fmt=3&VInst=PROD&VType=PQD&RQT=309&VName=PQD&TS=1245852022&clientId=5904 (accessed 4 November 2008).

19. K. Larry Storrs and Nina M. Serafino, "Andean Regional Initiative (ARI): FY2002 Supplemental and FY2003 Assistance for Colombia and Neighbors," Congressional Research Service Report, updated 22 January 2003, 31 and 42.

20. This replaced the Clinton administration's Presidential Decision Directive 73, which established a strict distinction between counterdrugs and counterinsurgency.

21. John A. Cope, "Colombia's War: Toward a New Strategy" (Washington, DC: National Defense University, Institute for National Strategic Studies, Strategic Forum 194), October 2002, 5–7.

22. Notes from a review of MILGP weekly reports and discussions with American military personnel.

23. Galen Jackman, "Media Roundtable with US Southern Command J-3 (Operations Chief)," 4 October 2002, http://ciponline.org/colombia/02100401.htm (accessed 3 June 2009).

24. Notes from a review of MILGP weekly reports.

25. Notes from discussion with Colombian civilian officials.

26. Republic of Colombia, "Colombia: Building a Path toward a New Horizon," Briefing, 2008.

27. Colombian Ministry of National Defense, "Logros de la Politicia de Consolidacion de La Seguridad Democratica—PCSD," Briefing, February 2009, http://www.mindefensa.gov.co/ (accessed 19 March 2009).

28. Jorge Enrique Mora Rangel, *Direccionamiento Estrategico, 2003*. Comando General Fuerzas Militares, 2003; William F. Perez, "An Effective Strategy for Colombia: A Potential End to the Current Crisis," US Army War College Student Paper, 3 May 2004, 10–11.

29. Carlos A. Ospina Ovalle, *Politicas de Commando: Guia Operacional 2003* (Bogotá: Fuerzas Militares de Colombia Ejercito Nacional, 2003).

30. US General Accounting Office, *Security Assistance: Efforts to Secure Colombia's Cano Limon-Covenas Oil Pipeline Have Reduced Attacks, but Challenges Remain* (Washington, DC: GAO, September 2005), 1–13.

31. Scott Wilson, "US Moves Closer to Colombia's War; Involvement of Special Forces Could Trigger New Wave of Guerrilla Violence," *Washington Post,* 7 February 2003, http://proquest.umi.com/pqdweb?index=3&did=2840683 41&SrchMode=2&sid=3&Fmt=3&VInst=PROD&VType=PQD&RQT=309&V Name=PQD&TS=1246038232&clientId=5904 (accessed 4 November 2008).

32. Notes from discussion with American military official.

33. Robert D. Kaplan, *Imperial Grunts: The American Military on the Ground* (New York, NY: Random House, 2005), 64 and 77.

34. US GAO, *Security Assistance: Efforts to Secure Colombia's Cano Limon-Covenas Oil Pipeline Have Reduced Attacks,* 3 and 15–18.

35. "1998 Bombing Cited as US Decertifies Unit in Colombia," *New York Times*, 15 January 2003, http://proquest.umi.com/pqdweb?index=1&did=27716 7311&SrchMode=2&sid=2&Fmt=3&VInst=PROD&VType=PQD&RQT=309& VName=PQD&TS=1246045418&clientId=5904 (accessed 4 November 2008); T. Christian Miller, "THE WORLD; Colombian Air Force Chief Quits; General Resigns Amid US Pressure and New Evidence Suggesting that Pilots Knowingly Fired on Civilians during a 1998 Bombing Raid," *Los Angeles Times*, 26 August 2003, http://proquest.umi.com/pqdweb?index=2&did=388820971&SrchMode=2 &sid=5&Fmt=3&VInst=PROD&VType=PQD&RQT=309&VName=PQD&TS =1246045550&clientId=5904 (accessed 4 November 2008).

36. Notes from a review of MILGP weekly reports.

37. Hill, "Statement before the House Armed Services Committee on the State of Special Operations Forces," 12 March 2003.

38. James T. Hill, "Colombia: Key to Security in the Western Hemisphere," The Heritage Foundation Lecture 790, 16 April 2003, http://www.heritage.org/ Research/LatinAmerica/HL790.cfm (accessed 15 September 2008).

39. Gabriel Marcella, *The United States and Colombia: The Journey from Ambiguity to Strategic Clarity* (Carlisle Barracks, PA: Strategic Studies Institute, May 2003), 21.

40. Notes from a review of MILGP weekly reports and discussions with American military officials.

41. "Colombian Government Issues Official Report on 'FARC Genocide,'" *BBC Monitoring Newsfile*, 6 May 2003, http://proquest.umi.com/pqdweb?index =0&did=332739061&SrchMode=2&sid=1&Fmt=3&VInst=PROD&VType= PQD&RQT=309&VName=PQD&TS=1246114047&clientId=5904 (accessed 4 November 2008); Margarita Martinez, "Colombian Chief Takes Blame in Rescue Fiasco, Sound of Helicopters Set Off an Orgy of Death in the Jungle," *San Antonio Express–News*, 7 May 2003, http://proquest.umi.com/pqdweb?index=3&did=780 326551&SrchMode=2&sid=2&Fmt=3&VInst=PROD&VType=PQD&RQT=309 &VName=PQD&TS=1246114784&clientId=5904 (accessed 4 November 2008).

42. "Colombia: Joint Operation Against FARC's Teofilo Forero Unit Bags 52 Guerillas," *BBC Monitoring Americas—Political*, 3 June 2003, http://proquest. umi.com/pqdweb?index=0&did=343349231&SrchMode=2&sid=3&Fmt=3& VInst=PROD&VType=PQD&RQT=309&VName=PQD&TS=1246115960& clientId=5904 (accessed 4 November 2008).

43. "Colombia: US Insists on Approval of Any Action to Free US Hostages," *BBC Monitoring Americas—Political*, 20 May 2003, http://proquest.umi.com/ pqdweb?index=0&did=338532281&SrchMode=2&sid=5&Fmt= 3&VInst=PRO D&VType=PQD&RQT=309&VName=PQD&TS=1246116365&clientId=5904 (accessed 4 November 2008).

44. Thomas A. Marks, "Colombian Military Support for 'Democratic Security,'" *Small Wars and Insurgencies* (June 2006): 203.

45. Republic of Colombia, *Democratic Security and Defense Policy*, 31–61.

46. Republic of Colombia, *Democratic Security and Defense Policy*, 37–39.

47. Notes from discussion with Colombian civilian official; Thomas A. Marks, *Sustainability of Colombian Military/Strategic Support for "Democratic Security"* (Carlisle Barracks, PA: Strategic Studies Institute, July 2005), 11–12; "Colombia: Peasant Soldiers Return Home after Completing Military Training," *BBC Monitoring Americas—Political*, 6 March 2003, http://proquest.umi.com/ pqdweb?index=0&did=301760851&SrchMode=2&sid=1&Fmt=3&VInst=PRO D&VType=PQD&RQT=309&VName=PQD&TS=1246135556&clientId=5904 (accessed 4 November 2008).

48. Rachel Van Dongen, "Colombia's Newest Troops Don't Have to Leave Home, Some 5,000 Troops Eagerly Enlist in a Program that Lets Them Serve in Own Villages," *Christian Science Monitor*, 9 April 2003, http://proquest.umi.com/ pqdweb?index=0&did=322434131&SrchMode=2&sid=2&Fmt=3&VInst=PRO D&VType=PQD&RQT=309&VName=PQD&TS=1246136585&clientId=5904 (accessed 4 November 2008).

49. Notes from discussions with Colombian and American civilian officials.

50. Marks, "Colombian Military Support for 'Democratic Security,'" 209; Notes from discussions with Colombian Army military officials.

51. Colombian Army, *Colombian Army Military History* (Bogotá: E3 Section, Army Historical Studies Center, 2007), 356–357; Notes from discussions with Colombian Military officials.

52. Marks, *Sustainability of Colombian Military/Strategic Support for "Democratic Security,"* 10–13.

53. Colombian Army, "Disposicion Numero 000021 Por medio de la cual se reorganiza el Ejercito Nacional," 29 September 2003. Document signed by Colombian Army Commanding General Carlos Alberto Ospina. This document reorganized the Colombian Army in accordance with Article 29 of Law 1512 of 11 August 2000.

54. Douglas Porch and Christopher W. Muller, "'Imperial Grunts' Revisited: The US Advisory Mission in Colombia," in *Military Advising and Assistance: From Mercenaries to Privatization, 1815–2007,* ed. Donald Stoker (London: Routledge, 2008), 175–176; Notes from review of MILGP weekly reports and discussions with American military officials.

55. Kenneth Finlayson, "OPATT to PATT: El Salvador to Colombian and the Formation of Planning and Assistance Training Teams," *Veritas, Journal of Army Special Operations History* 2, no. 4 (2006): 93.

56. Notes from discussions with American military officials. One official noted that the COLMIL commander did not request or formally approve the PATT concept. As he put it, the COLMIL commander did not say yes, but he did not say no.

57. Miller, "THE WORLD; Colombian Air Force Chief Quits."

58. Thomas C. Bruneau, "Restructuring Colombia's Defense Establishment to Improve Civilian Control and Military Effectiveness" (Monterey, CA: Center for Civil-Military Relations, Naval Postgraduate School (2004)), 4, http://www.resdal.org/experiencias/main-lasa-04.html (accessed 11 March 2009).

59. Andrew Selsky, "Colombian Military Commander Resigns," *Washington Post,* 13 November 2003, http://proquest.umi.com/pqdweb?index=7&did=44481 0391&SrchMode=2&sid=1&Fmt=3&VInst=PROD&VType=PQD&RQT=309& VName=PQD&TS=1246371401&clientId=5904 (accessed 4 November 2008); "Colombia Names New Armed Forces Leader," *New York Times,* 19 November 2003, http://proquest.umi.com/pqdweb?index=1&did=453199511&SrchMode=2 &sid=2&Fmt=3&VInst=PROD&VType=PQD&RQT=309&VName=PQD&TS =1246371594&clientId=5904 (accessed 4 November 2008).

60. "Army Chief Resolves to Get a Rebel Leader/Colombian General to Quit if He Fails," *Houston Chronicle,* 20 December 2003, http://proquest.umi.com/pqdweb?index=0&did=504360601&SrchMode=2&sid=4&Fmt=3&VInst=PRO D&VType=PQD&RQT=309&VName=PQD&TS=1246371940&clientId=5904 (accessed 4 November 2008).

61. "Colombia: Military Forces Chief Asks Generals to Locate, Rescue All Hostages," *BBC Monitoring Americas,* 27 November 2003, http://proquest. umi.com/pqdweb?index=0&did=465970841&SrchMode=2&sid=5&Fmt= 3&VInst=PROD&VType=PQD&RQT=309&VName=PQD&TS=124637 2576&clientId=5904 (accessed 4 November 2008).

62. "Army Chief Resolves to Get a Rebel Leader/Colombian General to Quit if He Fails."

63. US Department of State, "Colombia Country Report on Human Rights Practices 2003" (Washington, DC: US DOS, 25 February 2004), http://www. state.gov/g/drl/rls/hrrpt/2003/27891.htm (accessed 16 March 2009); US DOS, "International Narcotics Control Strategy Report for 2002."

64. Scott Wilson, "Colombia Targeting Rebel Strongholds; More Aggressive US-Backed Strategy Expected to Be More Challenging, Brutal," *Washington Post*, 25 January 2004, http://proquest.umi.com/pqdweb?index=4&did=530444221&Sr chMode=2&sid=1&Fmt=3&VInst=PROD&VType=PQD&RQT=309&VName=P QD&TS=1246389072&clientId=5904 (accessed 4 November 2008); In discussions with senior Colombian Military officials, several expressed a similar sentiment about the plight of the population in areas controlled by the narcoterrorists.

65. Notes from discussions with senior Colombian and American military officials.

66. Rachel Van Dongen, "Plan Puts Colombia on Offensive; Top US Officials Asked Congress Last Week to Increase the Cap on Troops Allowed in Colombia," *Christian Science Monitor*, 22 June 2004, http://proquest.umi.com/ pqdweb?index=0&did=653680301&SrchMode=2&sid=1&Fmt=3&VInst=PRO D&VType=PQD&RQT=309&VName=PQD&TS=1246456382&clientId=5904 (accessed 4 November 2008).

67. James T. Hill, "Statement before the House Armed Services Committee," 24 March 2004, http://armedservices.house.gov/comdocs/openingstatements andpressreleases/108thcongress/04-03-24hill.html (accessed 18 March 2009).

68. Hill, "Statement before the House Armed Services Committee."

69. Hill, "Statement before the House Armed Services Committee"; Benjamin Mixon, "Written Statement of Brigadier General Benjamin Mixon, United States Army, Director of Operations, United States Southern Command before the 108th Congress, Senate Armed Services Committee Subcommittee on Merging Threats and Capabilities," 2 April 2004, http://www.dod.mil/dodgc/olc/ docs/test04-04-02Mixon.doc (accessed 23 June 2009); Notes from a review of the weekly MILGP reports.

70. Hill, "Statement before the House Armed Services Committee."

71. Connie Veillette, "Andean Counterdrug Initiative (ACI) and Related Funding Programs: FY2005 Assistance," Congressional Research Service Report, updated 9 December 2004.

72. Alfred A. Valenzuela and Victor M. Rosello, "Expanding Roles and Missions in the War on Drugs and Terrorism: El Salvador and Colombia," *Military Review* (March–April 2004), 28–35.

73. US General Accounting Office, *Plan Colombia: Drug Reduction Goals Were Not Fully Met, but Security Has Improved; US Agencies Need More Detailed Plans for Reducing Assistance* (Washington, DC: GAO, October 2008).

74. Republic of Colombia, "CCAI: Centro de Coordinacion de Accion Integral," Briefing, 2007; Hill, "Statement before the House Armed Services Committee"; Notes from discussions with Colombian and American civilian and military officials.

75. US Department of State, Colombia Section, "International Narcotics Control Strategy Report for 2004" (Washington, DC: US DOS, March 2005), http://www.state.gov/p/inl/rls/nrcrpt/2005/vol1/html/42363.htm (accessed 1 October 2008).

76. "Colombia: Two Analysts See Serious Problems with Government's Security Policy," BBC Monitoring Americas, 20 April 2004, http://proquest.umi.com/pqdweb?index=0&did=621176981&SrchMode=2&sid=1&Fmt=3&VInst=PROD&VType=PQD&RQT=309&VName=PQD&TS=1246553296&clientId=5904 (accessed 4 November 2008).

77. Van Dongen, "Plan Puts Colombia on Offensive"; Edmund Turner, "Written Statement before the 109th Congress House Government Reform Committee Subcommittee on Criminal Justice, Drug Policy, and Human Resources," 10 May 2005, 6, http://www.dod.mil/dodgc/olc/docs/test05-05-10Turner.doc (accessed 19 March 2009).

78. US Department of State, "Colombia Country Report on Human Rights Practices 2004" (Washington, DC: US DOS, 28 February 2005), http://www.state.gov/g/drl/rls/hrrpt/2004/41754.htm (accessed 16 March 2009).

79. Garry Leech, "US/COLOMBIA: Demobilizing the AUC?" NACLA Report on the Americas, September/October 2004, http://proquest.umi.com/pqdweb?did=693199551&sid=2&Fmt=3&clientId=5094&RQT=309&VName=PQD (accessed 4 November 2008).

80. Kenneth Finlayson, "Colombian Special Operations Forces," Veritas, Journal of Army Special Operations History 2, no. 4 (2006): 56–59.

81. Manuel A. Orellana, "How to Train An Army of Intelligence Analysts" (Monterey, CA: Naval Postgraduate School Thesis, September 2005), 56.

82. Marks, Sustainability of Colombian Military/Strategic Support for "Democratic Security," 13.

83. Michelle L. Farrell, "Sequencing: Targeting Insurgents and Drugs in Colombia" (Monterey, CA: Naval Post Graduate School Thesis, March 2007), 74.

84. Ricardo A. Flores, "Improving the US Navy Riverine Capability: Lessons from the Colombian Experience" (Monterey, CA: Naval Postgraduate School Thesis, December 2007), 54.

85. Colombian MOD, "Logros de la Politicia de Consolidacion de La Seguridad Democratica—PCSD"; US DOS, "Colombia Country Report on Human Rights Practices 2004."

86. "FARC Offensive Signals Change in Tactics, Colombian Weekly Says," BBC Monitoring Americas, 14 February 2005, http://proquest.umi.com/pqdweb?index=0&did=792945081&SrchMode=2&sid=3&Fmt=3&VInst=PROD&VType=PQD&RQT=309&VName=PQD&TS=1246639601&clientId=5904 (accessed 4 November 2008); Farrell, "Sequencing: Targeting Insurgents and Drugs in Colombia," 77–78.

87. Andrew Feickert, "US Military Operations in the Global War on Terrorism: Afghanistan, Africa, the Philippines, and Colombia," Congressional Research Service Report, 26 August 2005, 18.

88. "FARC Attacks Seek to Discredit President, Colombian Weekly Says," *BBC Monitoring Americas*, 4 July 2005, http://proquest.umi.com/pqdweb?index =0&did=862530111&SrchMode=2&sid=1&Fmt=3&VInst=PROD&VType= PQD&RQT=309&VName=PQD&TS=1246639073&clientId=5904 (accessed 4 November 2008).

89. Indira A.R. Lakshmanan, "Rebels Reassert Deadly Agenda in Colombia," *Boston Globe*, 26 May 2005, http://proquest.umi.com/pqdweb?ind ex=2&did=844886731&SrchMode=2&sid=5&Fmt=3&VInst=PROD&VType =PQD&RQT=309&VName=PQD&TS=1246640478&clientId=5904 (accessed 4 November 2008).

90. Brantz J. Craddock, "Posture Statement before the 109th Congress House Armed Services Committee," 9 March 2005, http://ciponline.org/colombia/ 050309crad.pdf (accessed 1 October 2008).

91. Turner, "Written Statement before the 109th Congress House Government Reform Committee Subcommittee on Criminal Justice, Drug Policy, and Human Resources."

92. Notes from discussions with American military officials; Kenneth Finlayson, "'Conducting the Orchestra:' AOB 740 in Colombia," *Veritas, Journal of Army Special Operations History* 2, no. 4 (2006): 69–73; Robert W. Jones Jr., "'Who Taught Those Guys to Shoot Like Chuck Norris?' ODA 746 in Tolemaida," *Veritas, Journal of Army Special Operations History* 2, no. 4 (2006): 73–78; Kenneth Finlayson, "'There Is a Word I Need to Learn:' ODA 741 and Colombian National Police Training at Espinal," *Veritas, Journal of Army Special Operations History* 2, no. 4 (2006): 79–84; Robert W. Jones Jr., "Special Forces in Larandia: ODA 753 and the CERTE," *Veritas, Journal of Army Special Operations History* 2, no. 4 (2006): 85–90.

93. Notes from discussions with American military officials.

94. Finlayson, "OPATT to PATT," 92–93.

95. Notes from review of MILGP weekly reports and discussions with American and Colombian Military officials.

96. "Colombia: Armed Forces Restructured," *Stratfor Global Intelligence*, 5 May 2005, http://www.stratfor.com/memberships/63192/colombia_armed_ forces_restructured (accessed 2 July 2009).

97. Douglas Porch, "Uribe's Second Mandate, the War, and the Implications for Civil-Military Relations in Colombia," Naval Postgraduate School, Center for Contemporary Conflict, Strategic Insights 2 (February 2006), http://www.ccc.nps. navy.mil/si/2006/Feb/porchFeb06.asp (accessed 19 March 2009).

98. "Retired Generals Fear Consequences of Military Restructuring— Colombian Daily," *BBC Monitoring Americas,* 2 May 2005, http://proquest.umi. com/pqdweb?index=1&did=830857351&SrchMode=2&sid=1&Fmt=3&VInst= PROD&VType=PQD&RQT=309&VName=PQD&TS=1246652143&clientId= 5904 (accessed 4 November 2008).

99. Notes from discussions with senior Colombian and American civilian and military officials.

100. Notes from review of MILGP weekly reports and discussions with senior Colombian and American civilian and military officials.

101. Notes from discussions with senior Colombian and American civilian and military officials.

102. Porch and Muller, "'Imperial Grunts' Revisited: The US Advisory Mission in Colombia," 180–181.

103. Notes from discussions with former Colombian Army battalion commanders.

104. Notes from discussions with Colombian and American military officials; Porch and Muller, "'Imperial Grunts' Revisited: The US Advisory Mission in Colombia," 184–185.

105. Republic of Colombia, "Colombia: Building a Path toward A New Horizon," Briefing.

106. Steven C. Boraz, "Intelligence Reform in Colombia: Transparency and Effectiveness against Internal Threats," *Strategic Insights* (May 2007).

107. Notes from discussion with Colombian Military official.

108. US General Accounting Office, *Plan Colombia: Drug Reduction Goals Were Not Fully Met, but Security Has Improved,* 4–6.

109. US Department of State, "Report to Congress: US Assistance Program in Colombia and Plans to Transfer Responsibilities to Colombia," March 2006, http://ciponline.org/colombia/0603stat.pdf (accessed 1 October 2008).

110. Farrell, "Sequencing: Targeting Insurgents and Drugs in Colombia," 81.

111. US Department of State, "Colombia Country Report on Human Rights Practices 2005" (Washington, DC: US DOS, 8 March 2006), http://www.state.gov/g/drl/rls/hrrpt/2005/61721.htm (accessed 16 March 2009); US Department of State, Colombia Section, "International Narcotics Control Strategy Report for 2005" (Washington, DC: US DOS, March 2006), http://www.state.gov/p/inl/rls/nrcrpt/2006/vol1/html/62106.htm (accessed 1 October 2008).

112. Notes from discussions with Colombian and American military and civilian officials; Steven Dudley, "Colombian President Alienates Military Leaders," *McClatchey Tribune News Service*, 21 March 2006, http://proquest.umi.com/pqdweb?index=1&did=1007298581&SrchMode=2&sid=1&Fmt=3&VInst=PROD&VType=PQD&RQT=309&VName=PQD&TS=1246999653&clientId=5904 (accessed 4 November 2008).

113. Brantz J. Craddock, "Posture Statement before the 109th Congress House Armed Services Committee," 16 March 2006, http://ciponline.org/colombia/06031crad.pdf (accessed 1 October 2008).

114. Connie Veillette, "Colombia: Issues for Congress," Congressional Research Service Report, updated 4 January 2006, 10.

115. "Colombia Says Paramilitary Demobilization Complete," *BBC Monitoring Newsfile*, 18 April 2008, http://proquest.umi.com/pqdweb?index=0&did=1022593911&SrchMode=2&sid=1&Fmt=3&VInst=PROD&VType=PQD&RQT=309&VName=PQD&TS=1247072929&clientId=5904 (accessed 4 November 2008).

116. Jose R. Perdomo, "Colombia's Democratic Security and Defense Policy in the Demobilization of the Paramilitaries," US Army War College Student Paper, 29 March 2007, 15.

117. Farrell, "Sequencing: Targeting Insurgents and Drugs in Colombia," 99; Frechette, *Colombia and the United States—The Partnership,* 19.

118. "Democratic Security Policy to Continue, Says Colombia's New Defence Minister," *BBC Monitoring Americas*, 26 July 2006, http://proquest.umi.com/pq dweb?index=0&did=1083390701&SrchMode=2&sid=7&Fmt=3&VInst=PRO D&VType=PQD&RQT=309&VName=PQD&TS=1247081947&clientId=5904 (assessed 8 July 2009).

119. "Colombia Appoints New Head of Armed Forces," *Xinhua News Agency*, 16 August 2006, http://proquest.umi.com/pqdweb?index=1&did=109 5412221&SrchMode=2&sid=9&Fmt=3&VInst=PROD&VType=PQD&RQT= 309&VName=PQD&TS=1247083356&clientId=5904 (accessed 8 July 2009); Colombian Navy, *Closing the Gap: "Towards the Future"—The Naval Strategy, a Cornerstone in the Fight Against Narco-terrorism* (Bogotá: Colombian Navy, 2003).

120. "Colombia's New Armed Forces Chief 'Strategic' Intelligence Expert," *BBC Monitoring Americas*, 22 August 2006, http://proquest.umi.com/pqdweb?i ndex=0&did=1102991381&SrchMode=2&sid=1&Fmt=3&VInst=PROD&VTyp e=PQD&RQT=309&VName=PQD&TS=1247680055&clientId=5904 (accessed 4 November 2008).

121. "Colombian Public Losing Confidence in Military Intelligence, Daily Reports," *BBC Monitoring Americas*, 27 September 2006, http://proquest. umi.com/pqdweb?index=0&did=1136001071&SrchMode=2&sid=1&Fmt =3&VInst=PROD&VType=PQD&RQT=309&VName=PQD&TS=12471 55165&clientId=5904 (accessed 4 November 2008).

122. US Department of State, "Colombia Country Report on Human Rights Practices 2006" (Washington, DC: US DOS, 7 March 2007), http://www.state. gov/g/drl/rls/hrrpt/2006/78885.htm (accessed 16 March 2009).

123. "Commander Defends Colombia Armed Forces in Interview," *BBC Monitoring Americas*, 10 November 2006, http://proquest.umi.com/pqdweb?ind ex=0&did=1159878151&SrchMode=2&sid=2&Fmt=3&VInst=PROD&VType= PQD&RQT=309&VName=PQD&TS=1247155196&clientId=5904 (accessed 4 November 2008).

124. Colombian MOD, "Logros de la Politicia de Consolidacion de La Seguridad Democratica—PCSD"; Discussions with Colombian Military officials.

125. Paul Richfield, "Colombia Takes Its First Five Super Tucanos," *C4ISR Journal* (7 December 2006), http://www.c4isrjournal.com/story.php?F=2408477 (accessed 11 March 2009); "Colombia Buys 25 Super Tucanos," *Defesa@ Net*, 8 December 2005, http://www.defesanet.com.br/embraer/super_tucano_ colombia7_e.htm (accessed 11 March 2009); Notes from discussions with Colombian Military officials.

126. US Department of State, Colombia Section, "International Narcotics Control Strategy Report for 2006" (Washington, DC: US DOS, March 2007), http://www.state.gov/p/inl/rls/nrcrpt/2007/vol1/html/80855.htm (accessed 1 October 2008).

127. US DOS, "Colombia Report on Human Rights Practices 2006."

128. Colombian MOD, "Logros de la Politicia de Consolidacion de La Seguridad Democratica—PCSD."

129. Colombian Ministry of National Defense, *Politica de Consolidacion de La Seguridad Democratica (PCSD)* (Bogotá: Ministry of National Defense, 2007), http://www.mindefensa.gov.co/ (accessed 19 March 2009).

130. Colombian Ministry of National Defense, "Democratic Security Consolidation Policy," Briefing, 2008.

131. James Stavridis, "Posture Statement before the 110th Congress House Armed Services Committee," 21 March 2007, http://armedservices.house.gov/comdocs/schedules/2007.shtml (accessed 10 July 2009).

132. Notes from discussions with American civilian and military officials.

133. Notes from discussions with American and Colombian Military officials.

134. Constanza Vieira, "Colombia: Uribe Uses Military to Monitor Opposition Lawmakers," *Global Information Network*, 23 April 2007, http://proquest.umi.com/pqdweb?did=1259095951&sid=4&Fmt=3&clientId=5094&RQT=309&VName=PQD (accessed 3 November 2008).

135. "Colombian Military Links to Executions Could Hinder Human Rights Certification," *BBC Monitoring Americas*, 25 June 2007, http://proquest.umi.com/pqdweb?index=0&did=1294313041&SrchMode=2&sid=3&Fmt=3&VInst=PROD&VType=PQD&RQT=309&VName=PQD&TS=1247325082&clientId=5904 (accessed 3 November 2008).

136. "Colombian Military Chief Says Armed Forces Close to Defeating Guerrillas," *BBC Monitoring Americas*, 17 September 2007, http://proquest.umi.com/pqdweb?index=0&did=1337048011&SrchMode=2&sid=1&Fmt=3&VInst=PROD&VType=PQD&RQT=309&VName=PQD&TS=1247330956&clientId=5904 (accessed 3 November 2008).

137. Juan Forero, "Colombian Leader Says Rebels Killed 11 Civilian Hostages," *Washington Post*, 29 June 2007, http://proquest.umi.com/pqdweb?index=7&did=1296614111&SrchMode=2&sid=1&Fmt=3&VInst=PROD&VType=PQD&RQT=309&VName=PQD&TS=1247327973&clientId=5904 (accessed 3 November 2008).

138. Republic of Colombia, "CCAI: Centro de Coordinacion de Accion Integral," Briefing.

139. Juan Forero, "Colombia Tries Social Reform Programs as New Weapons Against Rebels; Aim Is to Establish a State Presence in Long-Neglected Areas," *Washington Post*, 10 July 2007, http://proquest.umi.com/pqdweb?index=0&did=1301875571&SrchMode=2&sid=1&Fmt=3&VInst=PROD&VType=PQD&RQT=309&VName=PQD&TS=1247333209&clientId=5904 (accessed 3 November 2008).

140. Peter DeShazo, et. al., *Colombia's Plan de Consolidacion Integral de la Macarena: An Assessment* (Washington, DC: Center for Strategic and International Studies, June 2009), 1–2, 7.

141. US Department of State, "Colombia Report on Human Rights Practices 2007" (Washington, DC: US DOS, 11 March 2008), http://www.state.gov/g/drl/rls/hrrpt/2007/100633.htm (accessed 16 March 2009); Notes from discussion with American military officials.

142. Republic of Colombia, "Colombia: Building a Path toward a New Horizon," Briefing.

143. Colombian Ministry of National Defense, "Consolidating Achievements," Briefing, July 2008, http://www.colombiaemb.org/docs/BRIEFING_COLOMBIA_SECURITY_JULY_2008.PPT (accessed 10 July 2009).

144. US Department of State, Colombia Section, "International Narcotics Control Strategy Report for 2007" (Washington, DC: US DOS, March 2008), http://www.state.gov/p/inl/rls/nrcrpt/2008/vol1/html/100776.htm (accessed 1 October 2008).

145. "Colombia's FARC Seen as Considerably Weakened But Not Defeated," *BBC Monitoring Americas*, 4 December 2007, http://proquest.umi.com/pqdweb?index=1&did=1392730681&SrchMode=2&sid=1&Fmt=3&VInst=PROD&VType=PQD&RQT=309&VName=PQD&TS=1247343082&clientId=5904 (accessed 3 November 2008).

146. "Colombian Military Plan Yet to Defeat FARC—Political Analyst," *BBC Monitoring Americas*, 20 January 2008, http://proquest.umi.com/pqdweb?did=1415609471&sid=3&Fmt=3&clientld=5094&RQT=309&VName=PQD (accessed 3 November 2008).

147. Juan Forero, "Colombian Rebel Commander Killed; Strike by Government Forces Called Major Setback for FARC Guerrillas," *Washington Post*, 2 March 2008, http://proquest.umi.com/pqdweb?index=17&did=1438193001&SrchMode=2&sid=1&Fmt=3&VInst=PROD&VType=PQD&RQT=309&VName=PQD&TS=1247497351&clientId=5904 (accessed 3 November 2008).

148. Hugh Bronstein, Joshua Goodman, and Bill Faries, "Top Rebel Commander Killed by His Own Bodyguard, Colombia Says; Second FARC Leader Killed as Countries Agree to Ease Tensions," *Ottawa Citizen*, 8 March 2008, http://proquest.umi.com/pqdweb?index=7&did=1443445081&SrchMode=2&sid=4&Fmt=3&VInst=PROD&VType=PQD&RQT=309&VName=PQD&TS=1247498664&clientId=5904 (accessed 3 November 2008).

149. Anonymous, "Manuel Marulanda," *Times*, 27 May 2008, http://proquest.umi.com/pqdweb?index=0&did=1484996861&SrchMode=2&sid=5&Fmt=3&VInst=PROD&VType=PQD&RQT=309&VName=PQD&TS=1247499810&clientId=5904 (accessed 3 November 2008).

150. James Stavridis, "US Southern Command 2008 Posture Statement," 13 March 2008, 15–16, http://www.southcom.mil/AppsSC/factFiles.php?id=43 (accessed 19 March 2009).

151. Notes from discussions with Colombian and American civilian and military officials.

152. US GAO, *Plan Colombia: Drug Reduction Goals Were Not Fully Met, but Security Has Improved,* 67.

153. US GAO, *Plan Colombia: Drug Reduction Goals Were Not Fully Met, but Security Has Improved,* 71.

154. US GAO, *Plan Colombia: Drug Reduction Goals Were Not Fully Met, but Security Has Improved;* 28, 47. Corrected addition errors in original.

155. Juan Forero, "Colombian Troops Kill Farmers, Pass Off Bodies as Rebels'," *Washington Post,* 30 March 2008, http://proquest.umi.com/pqdweb?index=2&did=1453809271&SrchMode=2&sid=1&Fmt=3&VInst=PROD&VType=PQD&RQT=309&VName=PQD&TS=1247514403&clientId=5904 (accessed 3 November 2008); Discussions with Colombian Military officers.

156. Jose de Cordoba, "Southern Front: Rebels Flail in Colombia after Death of Leader," *Wall Street Journal,* 28 May 2008, http://proquest.umi.com/pqdweb?did=1485594701&sid=1&Fmt=4&clientld5094&RQT=309&VName=PQD (accessed 3 November 2008).

157. John Otis, "How Colombian Army Hoodwinked the Rebels," *Houston Chronicle,* 4 July 2008, http://proquest.umi.com/pqdweb?index=4&did=1505420341&SrchMode=2&sid=1&Fmt=3&VInst=PROD&VType=PQD&RQT=309&VName=PQD&TS=1247584749&clientId=5904 (accessed 3 November 2008); Frank Bajak, "'Can This Truly Work?': Rebel Disarray, Acting Lessons, Payback All Part of Colombian Rescue," *North Adams Transcript,* 4 July 2008, http://proquest.umi.com/pqdweb?did=1505423661&sid=1&Fmt=3&clientld=5094&RQT=309&VName=PQD (accessed 3 November 2008).

158. Tony Allen-Mills, "Daring Sting Freed Jungle Hostages," *Sunday Times,* 6 July 2008, http://proquest.umi.com/pqdweb?did=1505859291&sid=6&Fmt=3&clientld=5094&RQT=309&VName=PQD (accessed 3 November 2008).

159. Juan Forero, "In Colombia Jungle Ruse, US Played a Quite Role; Ambassador Spotlights Years of Aid, Training," *Washington Post,* 9 July 2008, http://proquest.umi.com/pqdweb?index=3&did=1507416311&SrchMode=2&sid=3&Fmt=3&VInst=PROD&VType=PQD&RQT=309&VName=PQD&TS=1247586488&clientId=5904 (accessed 3 November 2008).

160. Juan Ferero, "Colombia Fires 27 From Army Over Killings; Youths' Deaths Attributed to Stress on Body Counts," *Washington Post,* 30 October 2008, http://proquest.umi.com/pqdweb?index=7&did=1585149411&sod=6&Fmt=3&clientld=5094&RQT=309&VName=PQD (accessed 3 November 2008).

161. "General Montoya's Replacement," *Semana.com,* 10 November 2008, http://www.semana.com/wf_ImprimirArticuloIngles.aspx?IdArt=117595 (accessed 15 January 2009).

162. Notes from discussion with Colombian and American civilian and military officials.

163. US Department of State, "Colombia Report on Human Rights Practices 2008" (Washington, DC: US DOS, 25 February 2009), http://www.state.gov/g/drl/rls/hrrpt/2008/wha/119153.htm (accessed 16 March 2009).

164. Colombian MOD, "Logros de la Politicia de Consolidacion de La Seguridad Democratica—PCSD," Briefing.

Chapter 4

Observations from the Colombian Security Force Experience

> Repeating an Afghanistan or an Iraq—forced regime change followed by nation-building under fire—probably is unlikely in the foreseeable future. What is likely though, even a certainty, is the need to work with and through local governments to avoid the next insurgency, to rescue the next failing state, or to head off the next humanitarian disaster.
>
> Correspondingly, the overall posture and thinking of the United States Armed Forces has shifted—away from solely focusing on direct American military action, and towards new capabilities to shape the security environment in ways that obviate the need for military intervention in the future.
>
> Secretary of Defense Robert M. Gates[1]

For many Americans the Colombian experience from 1998 to 2008 provides a successful example of working with and through local governments. In fact, some in government talk of a "Colombian model" that might have applicability—given the similar problems of illegal armed groups, ungoverned spaces, and drug trafficking—in places like Afghanistan or Mexico.[2] Although, when pressed, most have little idea—beyond generalities like long-term support, counternarcotics, and human rights—what that model might be. Then the United States "Plan Colombia" becomes a potential candidate.[3] However, with the passage of "expanded authorities" by Congress in 2002, "Plan Colombia," primarily the counterdrug component of President Andres Pastrana's nation-strengthening Plan Colombia, became the term for a modified US counterdrug program that permitted assistance to Colombia in its internal struggle against illegal armed groups with ties to narcotrafficking. In America's support of Uribe's Democratic Security and Defense Policy (DSDP), Colombian guerrillas became narcoterrorists. In truth, there was no great model or elaborate plan. Rather, a long-term trial-and-error process that included major policy changes, misunderstandings, frustrations, and mistakes by Americans and Colombians alike eventually produced the improvements in Colombian security that have led to progress in governance and counternarcotics efforts.

Reasons for Improvements in Colombian Security

Without a doubt, the leadership of President Alvaro Uribe constitutes the principal reason for the improvements in Colombia. First, unlike his predecessors, Uribe tackled the security problem as his number one priority. From the beginning, he intended to fight and defeat the illegal armed groups. He expanded the security forces and relentlessly pushed them to engage the narcoterrorists and to regain control of the countryside. Second, understanding that improved security was not an end in itself, but a prerequisite for pursuing other programs to address Colombia's ills, Uribe's DSDP and its follow-on Democratic Security Consolidation Policy (DSCP) established a multiyear, whole-of-government effort to establish governance and social programs throughout the country. Third, no one worked harder at pushing for results in multiple governmental areas than Uribe. He personally monitored and supervised the execution of his policies. After providing guidance and resources for various governmental programs, Uribe expected results. Fourth, as a strong, forceful leader who obtained positive results, Uribe retained the support of the Colombian people and the respect of the security forces. Willing—as necessary—to replace those who failed to meet his expectations, Uribe held his government officials accountable for their actions and for those of their subordinates. Last, despite the ups and downs of his presidency, the Colombians elected Uribe to a second term, which permitted a continuation of his policies for another 4 years. Without Uribe, the security and governance improvements in Colombia would not be what they are today.

The performance and leadership of the Colombian Military (COLMIL) constitutes the second major reason for the improvements in security. At the beginning of the Pastrana presidency, the COLMIL—particularly the Colombian Army—confronted the difficult task of combating large guerrilla units with minimal support from the government and with no US assistance. Even under Plan Colombia and after recertification, US assistance remained restricted to counterdrug programs. Under these trying conditions, the COLMIL leadership—primarily Tapias, Mora, and Ospina—developed an understanding of their threat, assessed the capabilities of their forces, and began a multiyear program for improvements within existing resources. Through professionalization, reorganization, better training, day-to-day improvements, and combat operations, the COLMIL blunted the guerrilla threat and developed an experience-based doctrine. Unlike other militaries that falter when provided additional resources and new missions, under President Uribe the COLMIL continued to build on its previous experience and reduced the areas under narcoterrorist control. In fact, under commanders like Ospina and Padilla, the COLMIL developed a military strategy, expanded its forces, increased its operations, refined its procedures, established specialized units—mountain battalions, antiterrorist, highway security, special operations, and *soldados campesinos*—and improved its human rights record while reducing the narcoterrorist capability. Despite its challenges, the COLMIL developed the capabilities that allowed it to seize and maintain the initiative against the narcoterrorists. To have come from the jaws of defeat in 1998 to the successes and capabilities of 2008 is a tribute to the vision, professionalism, and persistence of the senior COLMIL leadership and to the skills, dedication, and performance of security force personnel.

What about the role of US assistance? Many Americans immediately focus on the $6.5 billion spent—$2.1 billion from 1999 to 2002 and $4.4 billion from 2003 to 2008. (See Table 3, US assistance to Colombia, 1999–2002, and Table 5, US assistance to Colombia, 2003–2008.[4]) During the 4-year Pastrana period, the COLMIL received in noncounterdrug assistance no more than $10 million in training and equipment of which $6 million supported infrastructure security training for 18 Brigade in Arauca, $25 million for new antiterrorist units, and $73 million in nonlethal excess equipment—$108 million. Even with $47 million added for the counterdrug-funded riverine units, the COLMIL received only $155 million of the $2.1 billion. Basically, the COLMIL received no military assistance useful in its fight against the guerrillas. Things changed dramatically during the 6-year Uribe period with "expanded authorities," even though counterdrug funding totaled $2.4 of the $4.4 billion—more than half. The COLMIL received $1.3 billion for counterdrug activities—$48 million for Air Bridge Denial, $90 million for riverine units, and $1.2 billion for counterdrug operations that continued to include funding for the counterdrug brigade and its helicopter fleet and some Colombian Army operations against the narcoterrorists; and $970 million for the Colombian Army—$770 million for Foreign Military Sales (FMS) helicopters, $109 million for infrastructure security, and $88 million for training and equipping Army units. During this 6-year Uribe period, the COLMIL received over half of the assistance funds, but counterdrug funded programs did not always directly support its operations against the narcoterrorists. With US assistance, Uribe and his armed forces produced dramatic security improvements in Colombia. In contrast, the counterdrug programs—totaling over $4.4 billion during the 10 years—did not reduce coca cultivation.

With this funding, the US military provided assistance and support. This included suggestions, and at times requirements, for the security forces—human rights improvements, integrated national and military strategy, improved intelligence sharing, Ministry of National Defense (MOD) reform, improved jointness, intelligence-driven operations, special operations capability, and civil-military cooperation. Training for the security forces took many forms—individual, group, and unit—and addressed numerous topics: infantry skills, special operations, helicopter pilots and mechanics, combat life-saving, riverine operations, military decision making, intelligence, civil affairs, psychological operations, military law, and human rights to name a few. Human rights received the greatest emphasis. Human rights training and a focus on human rights programs improved the relations between the security forces and the populace. Training ranged from mobile training teams, planning assistance and training teams, specialists, and seminars in Colombia to individual training courses in the United States. Over time different priority units received instruction from US trainers—Antinarcotics Police Directorate (DIRAN), *Junglas*, riverine, counterdrug brigade, special operations, infrastructure security, Joint Task Force (JTF) *Omega*, *Carabineros*. Although the majority of security force personnel never worked with an American trainer, Colombian training and programs increasingly became US-influenced. In addition to training, US-funded equipment provided important capabilities—mobility from helicopters, riverine operations, antiterrorist units, and special operations. But perhaps most important, toward the end of this period US assistance programs made possible the sustainment of the COLMIL operations and of security force counterdrug programs that permitted the clearing, holding, and building of governance and social services throughout Colombia.

In contrast, while senior Colombian Military leaders and MOD officials acknowledged the importance of US assistance, several focused not on the amount of money or training or equipment—all of which they said were useful. Instead, they cited two things: the importance of being treated with professional respect and the value of interacting with US military personnel. The loss of its longstanding ties with the US military during decertification hit the COLMIL—particularly the Army—hard. A former COLMIL commander stated that the renewal and continuation of US assistance was very important for morale, but not as important for the military support provided. Another senior officer mentioned the importance of COLMIL being treated with dignity and respect.[5] Abandoned by its government and by the United States during decertification, renewed US ties provided the COLMIL an opportunity to regain public confidence and to demonstrate its professional competence. Being a partner with the US military raised morale and provided opportunities to work on problems together. Over the years, the COLMIL had been exposed to numerous US military personnel and to various US military concepts and suggestions—some considered useful, others not so. A former MOD official stated that US military assistance was critical—not, as he put it, for the little money—but for the "lots of knowledge" shared by American personnel when addressing Colombian problems.[6] Although not all the ideas worked or were accepted, the rigor and problem-solving approach of the Americans and their insistence on integrated planning, which helped in organizing and coordinating, improved Colombian security efforts. The impact of US assistance to the Colombian security forces cannot be measured just by the amount of money, training, equipment, or logistical support provided.

Insights on Working Together

The lack of a great model or an elaborate plan—things that rarely work if they really exist—does not mean that there are not important things to learn from the Colombian experience. In fact, the give-and-take nature of the process over 10 years offers valuable insights into the challenges that the US military faces when working, in the words of Secretary Gates, "with and through local governments." Much of what follows may seem basic or obvious; however, sometimes in the press of events and in the interactions with others it becomes difficult to understand or to accept the obvious.

● *It all begins with policy. US policy and Congressional actions provide the resources, constraints, and framework for what the United States will and will not attempt in a host nation, just as host nation policies, resources, constraints, and institutions determine what can and cannot be done.* The fundamental reason for the 30-year intensified involvement of the United States in Colombia is illegal drugs. In the beginning, US policy and funding focused on counterdrugs, on human rights, and on not being drawn into a counterinsurgency. The United States was in Colombia primarily to fight narcotrafficking, not to assist Colombia in other endeavors. After the terrorist attack on 9/11 and the collapse of Pastrana's peace negotiations in 2002, the United States agreed to support Colombia's fight against its long-term internal security threat—narcoterrorists. During this same period, the fundamental problem in Colombia had been security—not narcotics. Colombians elected Pastrana to negotiate a settlement with the guerrillas—the Revolutionary Armed Forces of Colombia (FARC) and the National Liberation Army (ELN). The COLMIL undertook a modest, resource-constrained reorganization and modernization program. After the failure of Pastrana's peace efforts, Colombia elected Uribe to defeat the guerrillas. Uribe inherited a small but improved security force and resources from Plan Colombia just when the United States agreed to partner in fighting the narcoterrorists. What the United States was willing to support and what Colombia was willing and able to do changed over time, but the security challenge remained the fundamental problem.

- *Host nation leadership matters.* Uribe provided a strong, energetic, and capable leadership to execute his DSDP and DSCP. The right man at the right time can make a difference, even in a government with weak institutions, in a society known for corruption, and in a country with ungoverned spaces. In addition, the leadership of his Armed Forces proved proactive, professional, loyal, and adaptable in addressing security requirements.

- *It is not easy. Tensions, misunderstandings, frustrations, and an adversarial relationship are normal when two nations work together.* Different interests, priorities, languages, cultures, systems, and procedures, among other things, mean that frustrations will always result from miscommunication, misunderstanding, confusion, distrust, and differing goals. As an agent of change and the controller of resources, the US role will tend to be seen as having—as often it does—"a heavy adversarial quotient."[7] Rather than accepting this situation as normal and working to mitigate the problem, often attitudes hardened to the point that both sides believed they knew best. A former US Ambassador to Colombia acknowledged that a major US error had been the failure to "understand the cultural differences" between Colombia and the United States.[8] Over time, both sides came to better understand the ways of the other.

- *The problem matters. Working with a host nation is more effective and less frustrating for all when its identification of the problem is similar to the US identification of the problem.* Before "expanded authorities" and Uribe, most Colombians defined their problem as security—guerrillas and narcotraffickers, and the United States defined its problem as narcotrafficking—illegal drugs. During the Pastrana presidency, the COLMIL focused on its security mission, defeating guerrillas, and resisted US pressure to being drawn into what the COLMIL considered the wrong effort, counterdrugs. To divert limited assets during a period of crisis from the critical task of combating the guerrillas made little sense to most in the COLMIL. During this period, Colombian–US military relations tended to be adversarial. After 2002, Colombia and the United States focused on combating the narcoterrorists—a security problem with links to narcotrafficking. Even so, the United States funded and conducted record numbers of counterdrug operations—extractions, interdictions, and extraditions. During this period, Colombian–US military relations improved, but could still be adversarial on certain issues.

- *Teamwork requires a team. Working* with *and* through *a host nation—listening to its concerns and understanding its perspective, its constraints, and its capabilities—is more effective and less frustrating for all than working* on *a host nation. This is particularly important in a support and assist role.* Working *with* and *through* a host nation—trying to make improvements to host nation forces, institutions, and procedures—is not the same as working *on* a host nation—trying to make fundamental changes in the host nation and its institutions. One way involves a partnership in which the host nation has the lead and the United States is in support. The US role is to make the host nation forces better than their opponents' forces. There will be disagreements over the best ways to accomplish this at times. Not only is working *with* and *through* more effective and less frustrating, it is more likely to produce sustainable host-nation solutions that will last. In fact, the best approach may be to improve, modify, or expand things that already work, rather than to create something new that no matter how perfect may not be feasible, acceptable, or sustainable by the host nation. The other way—working *on* a host nation by imposing unwanted solutions—is heavily adversarial, makes a long-term process longer, and often proves counterproductive. Before "expanded authorities," many in the COLMIL probably considered that they were being worked *on* given the US focus on counterdrugs, human rights vetting, and no other security support, rather than being worked *with* in addressing their number one priority—the security challenge.

- *Change is not simple. Modifying host nation institutions, organizations, procedures, and doctrine requires knowledge and understanding and is more effective and less frustrating—but more difficult and less common—that mirror-imaging or overlaying US organizations, procedures, and doctrine.* Individual and small-unit training is relatively easy and straightforward. Changing military institutions and organizations can be extremely difficult. Mirror imaging—making them look like us—is often the quick, easy, and common approach taken. It requires no special knowledge of the host nation military or its problems and it reinforces what some see as a "we-know-better, fix-it-quick" American approach. Unfortunately, to try to make the host nation like us even though its culture, society, government, and military are not like ours is a common thing to attempt, but a difficult thing to accomplish. To find ways to make the host nation military better at joint operations, intelligence sharing, military decision making, special operations, or noncommissioned leadership by building on its ways is often more productive than copying a foreign military organization. It is easy to overlook that the minimum standard for host nation forces is to be better than their opponent, not to become small versions of a world-class, high-tech military that the host nation cannot afford. Interestingly, the Colombian Army improved the training for its counterguerrilla battalions and mobile brigades—Colombian-developed units much smaller than their US counterparts—but made few other major organizational modifications to the fighting core of its forces. The COLMIL developed its strategic concepts, operational plans, and military units and conducted its operations—first securing its critical areas, then increasing its territorial control, and finally targeting specific high-value targets (HVTs)—with US support and assistance, but with Colombian forces and in a Colombian manner.

- *This is different. An understanding of the situations and the constraints faced by most host nation security forces—composed of military and national police organizations—in dealing with an internal threat should inform US counterinsurgency and stability operations doctrine.* Most countries facing internal threats have military and police forces engaged in this effort—something alien to US military experience, doctrine, and organization. As a result, many Americans did not know and had trouble understanding—much less accepting—the restrictions placed on the security forces by Colombian law. Operating under peacetime law and treating a combat site as a crime scene became a concept difficult for many to grasp, much less understand its implications for military operations. In many countries, the internal roles and missions of the Armed Forces and the national police are not well defined and the relationship between the two competitive at best, adversarial at worse. A basic question exists: Just how well suited can a modern, world-class military be to provide advice and support on internal conflicts for countries with an internally focused military and a national police force? Since the United States plans to work with and through host nation security forces in the near future, a better understanding of these forces and the environment in which they work becomes important.

- *Do not make things worse. US support to a limited portion of host nation security forces, such as military units with only counternarcotics missions and counternarcotics police, can distort security force roles and missions.* This clearly happened in Colombia. US emphasis on counterdrug programs—on the national police, on riverine units during decertification, and on a Colombian Army counterdrug brigade with its own helicopter fleet during "Plan Colombia"—exacerbated the rivalries between the police and the Armed Forces and between the military services. Critically limited resources, helicopters for example, ended up dispersed into five different organizations. Commenting on US support to the national police during the Pastrana period, a Colombian Army general told an American officer, "We now have two armies and you are responsible—you always support them—you treat them better than us!"[9] As a result, jointness and cooperation became more difficult when roles and missions became blurred.

● *Take the long view. Persistence—both by the host nation and by the United States—over a long time is required to deal with internal security problems.* Easily said, this proves difficult to do. For one reason, election cycles bring new leaders and new policies. For another, many problems require sustained policies and programs for multiple election cycles. Looking back in 2007 on Plan Colombia, Ambassador Myles R.R. Frechette identified another reason when he said, "The key US analytical error was not recognizing that Colombia's various problems took decades to develop and . . . [would] take at least a decade more of intense attention to approach resolution."[10] A 6-year timeline for accomplishing the nation-strengthening goals of Plan Colombia—even if it had been better focused, fully funded, and executed more effectively—proved unrealistic.

● *Smaller is better. A small footprint, limited resources, low profile, and some successes increase the likelihood for a long-term US involvement. In addition, a small footprint—limited military personnel—ensures the United States remains in a support role.* A veteran from El Salvador concluded, "Contrary to the US Defense Department's usual way of doing things, smaller is better." Staying small reinforced the facts: "It is their war and they must win it. . . . We probably cannot deliver victory from outside and if we can, it probably is transitory."[11] This observation remained valid in Colombia. Restrictions on Department of Defense (DOD) personnel may have made it difficult to execute support programs at times and the rotation of part-time personnel kept the workload heavy, but it helped ensure that the war remained a Colombian war, which is what the Colombians wanted. Looking back, one wonders if the effort would have been kept small if it had been in the area of responsibility of any other combatant command—one with a higher priority and greater resources than the United States Southern Command (USSOUTHCOM)—or if it had not been constrained by ongoing operations in other parts of the world. With higher priority and greater resources, one wonders if the effort would have stayed one-of-those-most-difficult missions for the US military—being in support of someone else. This may be one of the greatest challenges we face when working with host nations—resisting the temptation to do too much with too many. Better too few than too many.

This overview of the improvements in the Colombian security condition between 1998 and 2008 demonstrates that Colombians developed the key policies and did the heavy lifting. From 1998 to 2002, the COLMIL—with limited to no US assistance—developed a military strategy for dealing with its critical guerrilla situation and began a multiyear internal reorganization and professionalization project to execute that strategy. From 2002 to 2006, under Uribe's DSDP—which mobilized the government and most of the population—the security forces continued to grow in size, specialization, and effectiveness. COLMIL secured Bogotá and the key cities, attacked the FARC base-areas in southeast Colombia, and regained territorial control of much of Colombia. After 2006, the DSCP shifted the focus to a more whole-of-government effort to reestablish governance and social services in former narcoterrorist areas. Beginning in 2002, US support and assistance—logistics, aviation, intelligence, civil affairs, civil-military, special operations—permitted the COLMIL to execute its military strategy by supporting and sustaining operations throughout the countryside on a larger scale and for a longer time than possible without that assistance. Chairman of the Joint Chiefs of Staff, Admiral Mike Mullen, captured a key to the success in Colombia—and to doing it well with and through host nations in the future—when he emphasized the long-term United States support "to *your* approach, *your* execution, and obviously *your* results (emphasis added)."[12]

Notes

1.   Robert M. Gates, "US Global Leadership Campaign," Speech delivered 15 July 2008, http://www.defenselink.mil/utility/printitem.aspx?print=http://www.defenselink.mil/speeches/speech.aspx?speechid=1262 (accessed 15 July 2009). Gates' quotation is used as the introduction to the second chapter, "Stability in Full Spectrum Operations," of FM 3-07, *Stability Operations* (Washington, DC: Department of the Army, 6 October 2008), 2–1.

2.   Chairman of the Joint Chiefs of Staff, "Press Conference with Chairman of the Joint Chiefs of Staff Admiral Michael G. Mullen, Colombian Minister of Defense Juan Manuel Santos, and Colombian Military Forces Commander General Freddy Padilla," 5 March 2009, http://www.jcs.mil/speech.aspx?ID=1139 (accessed 20 July 2009); Scott Wilson, "Which Way in Afghanistan? Ask Colombia for Directions," *Washington Post*, 5 April 2009, http://www.washingtonpost.com/wp-dyn/content/article/2009/04/03/AR2009040302135_pf.html (accessed 6 April 2009).

3.   As throughout this study, "Plan Colombia" refers to the US program created initially to support a part of President Pastrana's Plan Colombia and which was modified by "expanded authorities" in 2002.

4.   Nina M. Serafino, "Colombia: Summary and Tables on US Assistance, FY1989–FY2003," Congressional Research Service Report, 3 May 2002, 4; US General Accounting Office, *Plan Colombia: Drug Reduction Goals Were Not Fully Met, but Security Has Improved; US Agencies Need More Detailed Plans for Reducing Assistance* (Washington, DC: GAO, October 2008), 28, 47.

5.   Notes from discussions with senior Colombian Military officials.

6.   Notes from discussion with senior Colombian civilian official.

7.   John D. Waghelstein, "Military-to-Military Contacts: Personal Observations—The El Salvador Case," *Low Intensity Conflict & Law Enforcement* (Summer 200): 39.

8.   Myles R.R. Frechette, *Colombia and the United States—The Partnership: But What is the Endgame?* (Carlisle Barracks, PA: Strategic Studies Institute, February 2007), 28.

9.   Notes from discussion with American military official.

10.   Frechette, *Colombia and the United States—The Partnership,* 27.

11.   Waghelstein, "Military-to-Military Contacts: Personal Observations—The El Salvador Case." 39.

12.   Chairman of the JCS, "Press Conference with Mullen, Santos, and Padilla." Emphasis added.

Glossary

AAR	after action review
ACI	Andean Counterdrug Initiative
ACOEA	Air Force Special Forces Group
AFEAU	*Agrupación de Fuerzas Especiales Antiterroristas Urbanas* (Joint Urban Special Antiterrorist Forces Group)
AFEUR	*Agrupación de Fuerzas Especiales Urbanas* (Army Urban Special Forces Group)
AFSOUTH	US Southern Air Force
ARI	Andean Regional Initiative
AUC	United Self-Defense Groups of Colombia
BCG	counterguerrilla battalions
BRCNA	*Brigada Contra el Narcotrafico* (Counterdrug Brigade)
BRIM	mobile infantry brigade
CACOM	Air Combat Command
CCAI	Coordination Center for Integrated Action
CCC	Joint Command No. 1—Caribbean
CCOPE	Joint Special Operations Command
CD	counterdrug
CDTS	counterdrug training support
CERTE	Army Tactical Retraining Center
CIA	Central Intelligence Agency
CIME	Army Military Intelligence Center
CNP	Colombian National Police
COESE	Army Special Operations Command
COLAF	Colombian Air Force
COLAR	Colombian Army
COLMIL	Colombian Military (Armed Forces)
COLNAV	Colombian Navy
CONVIVIR	Community Associations of Rural Vigilance
DAS	Department of Administrative Security
DEA	Drug Enforcement Agency
DIA	Defense Intelligence Agency
DIRAN	Antinarcotics Police Directorate
DMZ	demilitarized zone
DOD	Department of Defense
DOJ	Department of Justice
DOS	Department of State
DSCP	Democratic Security Consolidation Policy
DSDP	Democratic Security and Defense Policy
ELN	National Liberation Army

EP	People's Army
FARC	Revolutionary Armed Forces of Colombia
FARC-EP	Revolutionary Armed Forces of Colombia–People's Army
FBI	Federal Bureau of Investigation
FL	Florida
FLIR	forward looking infrared radar
FMAP	Foreign Military Assistance Program
FMF	Foreign Military Financing
FMS	Foreign Military Sales
FRAGO	fragmentary order
FTO	Foreign Terrorist Organization
FUDRA	*Fuerza de Desplique Rapido* (Rapid Reaction Force)
FURAD	*Fuerza de Accion Decisiva* (Decisive Action Force)
FURED	division reaction force
FY	fiscal year
GA	Georgia
GAO	General Accounting Office (effective 7 July 2004 the legal name became Government Accountability Office)
GAULA	Groups of Action Unified for the Liberation of Persons
GDP	gross domestic product
HVT	high-value target
IED	improvised explosive device
IMET	International Military Education and Training
INCLE	International Narcotics Control Law Enforcement
ISS	COLMIL Infrastructure Security
JAG	Judge Advocate General
JCET	Joint Combined Exchange Training
JIATF-S	Joint Interagency Task Force–South
JIC	joint intelligence center
JOC	joint operations center
JOC-S	joint operations center–south
JPAT	Joint Planning and Assistance Team
JTF	joint task force
JTF-S	Joint Task Force–South
MARFORSOUTH	Marine Force South
MILGP	Military Group
MOD	Ministry of National Defense
MTT	mobile training team
NADR	Non-proliferation, Anti-terrorism, Demining, and Related Programs
NAS	Narcotics Affairs Section
NC	North Carolina

NCO	noncommissioned officer
NGO	nongovernmental organization
NSPD	National Security Presidential Directive
ODA	Operational Detachment Alpha
OPATT	Operational Planning Assistance and Training Team
PATT	Planning and Assistance Training Team
PCCP	Plan Colombia Consolidation Phase
PCIM	*Plan de Consolidacion Integral de la Macarena*
PEEV	*Plan Especial Energetico Vial*
POW	prisoner of war
RCE	riverine combat elements
RIME	Army Military Intelligence Regions
SF	Special Forces
TDY	temporary duty
UN	United Nations
US	United States
USAID	US Agency for International Development
USARSO	US Army South
USCG	US Coast Guard
USNAVSO	US Navy South
USSOCSOUTH	US Special Operations Command South
USSOUTHCOM	United States Southern Command
VMOD	vice-minister of defense
WHINSEC	Western Hemisphere Institute for Security Cooperation

Bibliography

Primary Sources

Amnesty International. *"Leave Us in Peace!" Targeting Civilians in Colombia's Internal Armed Conflict.* London: Amnesty International, 2008. http://www.hrw.org/en/reports/2008/10/16/breaking-grip (accessed 18 March 2008).

Bonett Locarno, Manuel Jose. *Estrategia General De Las Fuerzas Militares" "Por la seguirdad de la poblacion y sus recurrsos."* Fuerzas Militares De Colombia Comando General, December 1997.

Carenno, Martin Orlando. *Direccionamiento 2004: Politicas de Comando.* Bogotá: Fuerzas Militares de Colombia Ejercito Nacional, 2004.

Chairman of the Joint Chiefs of Staff. "Press Conference with Chairman of the Joint Chiefs of Staff Admiral Michael G. Mullen, Colombian Minister of Defense Juan Manuel Santos, and Colombian Military Forces Commander General Freddy Padilla" (5 March 2009). http://www.jcs.mil/speech.aspx?ID=1139 (accessed 20 July 2009).

Colombian Army. *Colombian Army Military History.* Bogotá: E3 Section, Army Historical Studies Center, 2007.

———. "Disposicion Numero 000011 Por medio de la cual se reorganiza el Ejercito Nacional," 23 August 1996.

———. "Disposicion Numero 000010 Por medio de la cual se reorganiza el Ejercito Nacional," 23 November 1999.

———. "Disposicion Numero 000002 Por medio de la cual se reorganiza el Ejercito Nacional," 4 February 2002.

———. "Disposicion Numero 000021 Por medio de la cual se reorganiza el Ejercito Nacional," 29 September 2003.

Colombian Ministry of National Defense. "Consolidating Achievements." Briefing, July 2008. http://www.colombiaemb.org/docs/BRIEFING_COLOMBIA_SECURITY_JULY_2008.PPT (accessed 10 July 2009).

———. "Democratic Security Consolidation Policy." Briefing, 2008.

———. "Logros de la Politicia de Consolidacion de La Seguridad Democratica—PCSD." Briefing, February 2009. http://www.mindefensa.gov.co/ (accessed 19 March 2009).

———. *Politica de Consolidacion de La Seguridad Democratica (PCSD).* Bogotá: Ministry of National Defense, 2007. http://www.mindefensa.gov.co/ (accessed 19 March 2009).

Colombian Navy. *Armanda Nacional: Politicas Institucionales.* Bogotá: Colombian Navy, 2006.

———. *Closing the Gap: "Towards the Future"—The Naval Strategy, A Cornerstone in the Fight Against Narco-terrorism.* Bogotá: Colombian Navy, 2003.

———. *Closing the Gap: Towards the Future—The Naval Strategy Against Narco-terrorism.* Bogotá: Colombian Navy, 2008.

Craddock, Brantz J. "Posture Statement before the 109th Congress House Armed Services Committee," 9 March 2005. http://ciponline.org/colombia/050309crad.pdf (accessed 1 October 2008).

―――. "Posture Statement before the 109th Congress House Armed Services Committee," 16 March 2006. http://ciponline.org/colombia/06031crad.pdf (accessed 1 October 2008).

Cronin, Audrey K. "The 'FTO List' and Congress: Sanctioning Designated Foreign Terrorist Organizations." Congressional Research Service Report, 21 October 2003.

Feickert, Andrew. "US Military Operations in the Global War on Terrorism: Afghanistan, Africa, the Philippines, and Colombia." Congressional Research Service Report, 26 August 2005.

Grossman, Marc. "Testimony of Ambassador Marc Grossman, Under Secretary of State for Political Affairs, before the House Appropriations Committee's Subcommittee on Foreign Operations on US Assistance to Colombia and the Andean Region," 10 April 2002. http://www.ciponline.org/colombia/02041001.htm (accessed 29 May 2009).

Hill, James T. "Colombia: Key to Security in the Western Hemisphere." The Heritage Foundation Lecture 790, 16 April 2003. http://www.heritage.org/Research/LatinAmerica/HL790.cfm (accessed 15 September 2008).

―――. "Statement before the House Armed Services Committee on the State of Special Operations Forces," 12 March 2003. http://armedservices.house.gov/comdocs/openingstatementsandpressreleases/108thcongress/03-03-12hill.html (accessed 18 March 2009).

―――. "Statement before the House Armed Services Committee," 24 March 2004. http://armedservices.house.gov/comdocs/openingstatementsandpressreleases/108thcongress/04-03-24hill.html (accessed 18 March 2009).

Human Rights Watch. *Colombia's Killer Networks: The Military-Paramilitary Partnership and the United States*. New York, NY: Human Rights Watch, 1 November 1996. http://www.hrw.org/en/reports/1996/11/01/colombia-s-killer-networks?print (accessed 22 January 2009).

―――. *The Ties That Bind: Colombia and Military-Paramilitary Links*. New York, NY: Human Rights Watch, 1 February 2000. http://www.hrw.org/en/reports/2000/02/01/ties-bind?print (accessed 23 January 2009).

―――. *State of War: Political Violence and Counterinsurgency in Colombia*. New York, NY: Human Rights Watch, December 1993. http://www.hrw.org/legacy/reports/1993/colombia/statetoc.htm (accessed 26 January 2009).

―――. *The "Sixth Division": Military-Paramilitary Ties and US Policy in Colombia*. New York, NY: Human Rights Watch, September 2001. http://www.colombianatverket.se/files/thesixthdivision_hrw.pdf (accessed 8 March 2009).

Jackman, Galen. "Media Roundtable with US Southern Command J-3 (Operations Chief)," 4 October 2002. http://ciponline.org/colombia/02100401.htm (accessed 3 June 2009).

McCaffrey, Barry. "After Action Report—Visit Colombia 23–26 September 2007." E-mail Memorandum for USSOUTHCOM Commander, 29 September 2007.

Mixon, Benjamin. "Written Statement of Brigadier General Benjamin Mixon, United States Army, Director of Operations, United States Southern Command before the 108th Congress, Senate Armed Services Committee, Subcommittee on Emerging Threats and Capabilities," 2 April 2004. http://www.dod.mil/dodgc/olc/docs/test04-04-02Mixon.doc (accessed 23 June 2009).

Montoya Uribe, Mario. "Ejercito Nacional: Informacion del Comandante." Briefing, May 2006.

Mora Rangel, Jorge Enrique. *Direccionamiento Estrategico, 2003*. Comando General Fuerzas Militares.

———. *Guia de Planeamiento Estrategico, 2001*. Fuerzas Militares de Colombia Ejercito Nacional, 2001.

———. "Military Review's Interview with General Jorge Enrique Mora Rangel, Commander, Colombian Army." *Military Review* (March–April 2002): 33–37.

Ospina Ovalle, Carlos A. "Insights from Colombia's 'Prolonged War.'" *Joint Forces Quarterly* (Third Quarter, 2006): 57–61.

———. *Politicas de Commando: Guia Operacional 2003*. Bogotá: Fuerzas Militares de Colombia Ejercito Nacional, 2003.

———. "12 Questions for Gen. (Ret.) Carlos Ospina Ovalle," 23 March 2007. http://www.ndu.edu/chds/chdshome (accessed 8 July 2009).

———. "The Defeat of the FARC." Washington, DC: Center for Hemispheric Defense Studies Regional Insights #5, 15 March 2008. http://www.ndu.edu/chds/index.cfm?lang=EN&pageID=110&type=page (accessed 1 October 2008).

Pace, Peter. "Statement before the House Appropriations Defense Subcommittee," 4 April 2001. http://www.armed-services.senate.gov/statemnt/2001/011025pace.pdf (accessed 8 March 2009).

Republic of Colombia. "CCAI: Centro de Coordinacion de Accion Integral." Briefing, 2007.

———. "Colombia: Building a Path toward a New Horizon." Briefing, 2008.

———. *Democratic Security and Defense Policy*. Republic of Colombia, 16 June 2003.

———. "Plan Colombia: Plan for Peace, Prosperity, and the Strengthening of the State," October 1999. http://www/ciponline.org/colombia/plancolombia.htm (accessed 4 August 2008).

———. "Plan Colombia: Plan for Peace, Prosperity, and the Strengthening of the State." US Institute of Peace, posted 15 May 2000. http://www.usip.org/library/pa/colombia/adddoc/plan_colombia_101999.html (accessed 18 May 2009).

———. "Text of the Constitution of Colombia (1991)." http://confinder.richmond.edu/admin/docs/colombia_const2.pdf (accessed 13 April 2009).

Rodman, Peter. "Testimony before House International Relations Committee Subcommittee for Western Hemisphere," 11 April 2002. http://ciponline. org/colombia/02041104.htm (accessed 15 October 2008).

Serafino, Nina M. "Colombia: Conditions and US Policy Options." Congressional Research Service Report. Washington, DC: Library of Congress, 2001.

———. "Colombia: Summary and Tables on US Assistance, FY1989–FY2003." Congressional Research Service Report, 3 May 2002.

———. "Colombia: The Problem of Illegal Narcotics and US-Colombian Relations." Congressional Research Service Report, updated 11 May 1998.

———. "Colombia: The Uribe Administration and Congressional Concerns." Congressional Research Service Report, 14 June 2002.

———. "Colombia: US Assistance and Current Legislation." Congressional Research Service Report, updated 15 May 2001.

Sheridan, Brian E. "Statement before House Armed Services Committee," 23 March 2000. http://armedservices.house.gov/comdocs/testimony/ 106thcongress/00-03-23sheridan.htm (accessed 15 October 2008).

———. "Statement before House Committee on International Relations Subcommittee on Western Hemisphere," 21 September 2000.

Speer, Gary D. "Posture Statement of Major General Gary D. Speer, United States Army Acting Commander in Chief United States Southern Command before 107th Congress Senate Armed Services Committee," 5 March 2002. http://www.ciponoline.org/colombia/02030501.htm (accessed 1 October 2008).

Storrs, K. Larry, and Nina M. Serafino. "Andean Regional Initiative (ARI): FY2002 Assistance for Colombia and Neighbors." Congressional Research Service Report, updated 21 December 2001.

Stavridis, James. "Posture Statement before the 110th Congress House Armed Services Committee," 21 March 2007.

———. "US Southern Command 2008 Posture Statement," 13 March 2008. http:// www.southcom.mil/AppsSC/factFiles.php?id=43 (accessed 19 March 2009).

———. "Andean Regional Initiative (ARI): FY2002 Supplemental and FY2003 Assistance for Colombia and Neighbors." Congressional Research Service Report, updated 22 January 2003.

Turner, Edmund. "Written Statement before the 109th Congress House Government Reform Committee Subcommittee on Criminal Justice, Drug Policy, and Human Resources," 10 May 2005. http://www.dod.mil/dodgc/olc/docs/ test05-05-10Turner.doc (accessed 19 March 2009).

United Nations. "Report by the United Nations High Commissioner for Human Rights." United Nations Commission on Human Rights, 9 March 1998. http://www.unhchr.ch/Huridocda/Huridoca.nsf/fb00da486703f751c125 65a90059a227/e84a158767e2e9ecc12566180049cd3e?OpenDocument (accessed 21 April 2009).

———. "Report of the United Nations High Commissioner for Human Rights on the Human Rights Situation in Colombia." United Nations Commission on Human Rights, 24 February 2003. http://www.unhchr.ch/Huridocda/Huridoca.nsf/0/1304674285b7eb3bc1256cf5003906fb?Opendocument (accessed 19 March 2009).

———. "Report of the United Nations High Commissioner for Human Rights on the Situation of Human Rights in Colombia." United Nations General Assembly Human Rights Council, 29 February 2008. http://cc.msnscache.com/cache.aspx?q=%22report+of+the+united+nations+high+commissioner+for+human+rights%22+colombia+29+february+2008&d=75662309408647&mkt=en-US&setlang=en-US&w=a81b0a50,703c7d27 (accessed 19 March 2009).

US Department of State. "Colombia Country Report on Human Rights Practices." Washington, DC: US DOS, 1996–1998. Replace xxxx with year in the following Web address: http://www.state.gov/www/global/human_rights/xxxx_hrp_report/colombia.html. From 1999 to present: http://www.state.gov/g/drl/rls/ hrrpt/index.htm (accessed 16 March 2009).

———. "Foreign Military Training and DOD Engagement Activities of Interest." Washington, DC: US DOS, 1999–2000. http://www.state.gov/www/global/arms/fmtrain/toc.html. From 2001 to 2007: http://www.state. gov/t/pm/rls/rpt/fmtrpt/index.htm (accessed 1 October 2008).

———. "International Narcotics Control Strategy Reports." Washington, DC: US DOS, 1996–2008. http://www.state.gov/p/inl/rls/nrcrpt/ (accessed 1 October 2008).

———. "International Narcotics Control Strategy Report for 2002." Washington, DC: US DOS, 1 March 2003. http://www.state.gov/p/inl/rls/nrcrpt/2002/index.htm (accessed 1 October 2008).

———. "Report to Congress: US Assistance Program in Colombia and Plans to Transfer Responsibilities to Colombia," March 2006. http://ciponline.org/colombia/0603stat.pdf (accessed 1 October 2008).

US General Accounting Office. "Briefing Paper: Military Assistance Under Plan Colombia is Substantially Behind Schedule," June 2002. http://ciponline.org/colombia/02060001.htm (accessed 1 October 2008).

———. *Drug Control: Challenges in Implementing Plan Colombia.* Washington, DC: GAO, 12 October 2000.

———. *Drug Control: Financial and Management Challenges Continue to Complicate Efforts to Reduce Illicit Drug Activities in Colombia.* Washington, DC: GAO, 3 June 2003.

———. *Drug Control: Narcotics Threat From Colombia Continues to Grow.* Washington, DC: GAO, June 1999.

———. *Drug Control: Specific Performance Measures and Long-Term Costs for US Programs in Colombia Have Not Been Developed.* Washington, DC: GAO, June 2003.

———. *Drug Control: US Assistance to Colombia Will Take Years to Produce Results*. Washington, DC: GAO, October 2000.

———. *Drug Control: US Counternarcotics Efforts in Colombia Face Continuing Challenges*. Washington, DC: GAO, February 1998.

———. *Drug Control: US Nonmilitary Assistance to Colombia is Beginning to Show Intended Results, But Programs Are Not Readily Sustainable*. Washington, DC: GAO, July 2004.

———. *Military Training: Management and Oversight of Joint Combined Exchange Training*. Washington, DC: GAO, July 1999.

———. *Plan Colombia: Drug Reduction Goals Were Not Fully Met, but Security Has Improved; US Agencies Need More Detailed Plans for Reducing Assistance*. Washington, DC: GAO, October 2008.

———. *School of the Americas: US Military Training for Latin American Countries*. Washington, DC: GAO, August 1996.

———. *Security Assistance: Efforts to Secure Colombia's Cano Limon-Covenas Oil Pipeline Have Reduced Attacks, but Challenges Remain*. Washington, DC: GAO, September 2005.

US House of Representatives. *Transcript, Hearing of the House Committee on International Relations on US Anti-Drug Policy Towards Colombia*, 31 March 1998. http://www.colombiasupport.net/199803/hr033198.html (accessed 15 October 2008).

Veillette, Connie. "Andean Counterdrug Initiative (ACI) and Related Funding Programs: FY2005 Assistance." Congressional Research Service Report, updated 9 December 2004.

———. "Colombia: Issues for Congress." Congressional Research Service Report, updated 4 January 2006.

———. "Plan Colombia: A Progress Report." Congressional Research Service Report, updated 17 February 2005.

———. "Plan Colombia: A Progress Report." Congressional Research Service Report, updated 22 June 2005.

———. "Plan Colombia: A Progress Report," Congressional Research Service Report, updated 11 January 2006.

White House. "Report on US Policy and Strategy Regarding Counterdrug Assistance to Colombia and Neighboring Countries," 26 October 2000. Center for International Policy's Colombia Program. http://ciponline.org/colombia/102601.htm (accessed 1 October 2008).

Wilhelm, Charles E. "Statement before the 105th Congress Committee on Armed Services, United States Senate," 5 March 1998. http://www.fas.org/irp/congress/1998_hr/s980305w.htm (accessed 15 October 2008).

———. "Statement before the Senate Armed Services Committee," 13 March 1999. http://www.armed-services.senate.gov/statemnt/1999/990413cw.pdf (accessed 1 October 2008).

———. "Statement before the Senate Armed Services Committee," 7 March 2000. http://www.armed-services.senate.gov/statemnt/2000/000307cw.pdf (accessed 1 October 2008).

Secondary Sources

"1998 Bombing Cited as US Decertifies Unit in Colombia." *New York Times*, 15 January 2003. http://proquest.umi.com/pqdweb?index=1&did=2771 67311&SrchMode=2&sid=2&Fmt=3&VInst=PROD&VType=PQD& RQT=309&VName=PQD&TS=1246045418&clientId=5904 (accessed 4 November 2008).

Adams, David. "President Declares Emergency as 100 Die in Colombia." *Times,* 13 August 2002. http://proquest.umi.com/pqdweb?index=13&did=148 748841&SrchMode=2&sid=1&Fmt=3&VInst=PROD&VType=PQD& RQT=309&VName=PQD&TS=1245765801&clientId=5904 (accessed 4 November 2008).

Adams, David, and Paul de la Garza. "Contract's End Hints of Colombia Trouble." *St. Petersburg Times,* 13 May 2001. http://www.sptimes.com/ News/051301/news_pf/Worldandnation/Contract_s_end_hints_.shtml (accessed 4 March 2009).

Allen-Mills, Tony. "Daring Sting Freed Jungle Hostages," *Sunday Times*, 6 July 2008, http://proquest.umi.com/pqdweb?did=1505859291&sid=6&Fmt =3&clientld=5094&RQT=309&VName=PQD (accessed 3 November 2008).

Anonymous. "Manuel Marulanda." *Times*, 27 May 2008. http://proquest.umi. com/pqdweb?index=0&did=1484996861&SrchMode=2&sid=5&Fmt= 3&VInst=PROD&VType=PQD&RQT=309&VName=PQD&TS=1247 499810&clientId=5904 (accessed 3 November 2008).

"Another Beating." *The Economist* (10 January 1998). http://proquest.umi.com/ pqdweb?did= 25330049&sid=4&Fmt=3&clientid=5904&RQT=309&V Name=PQD (accessed 5 November 2008).

"Army Chief Resolves to Get a Rebel Leader/Colombian General to Quit if He Fails." *Houston Chronicle*, 20 December 2003. http://proquest.umi.com/ pqdweb?index=0&did=504360601&SrchMode=2&sid=4&Fmt=3&V Inst=PROD&VType=PQD&RQT=309&VName=PQD&TS=124637 1940&clientId=5904 (accessed 4 November 2008).

Aviles, William. "Paramilitarism and Colombia's Low-Intensity Democracy." *Journal of Latin American Studies* 38 (2006): 379–408.

Bajak, Frank. "'Can This Truly Work?': Rebel Disarray, Acting Lessons, Payback All Part of Colombian Rescue." *North Adams Transcript*, 4 July 2008. http://proquest.umi.com/pqdweb?did=1505423661&sid=1&Fmt=3&cli entld=5094&RQT=309&VName=PQD (accessed 3 November 2008).

———. "Colombia's New President Purges Military Leadership." *San Antonio Express–News,* 10 August 1998. http://proquest.umi.com/ pqdweb?index=7&did=1206820531&SrchMode=2&sid=4&Fmt=3& VInst=PROD&VType=PQD&RQT=309&VName=PQD&TS=12420 63078&clientId=5904 (accessed 11 May 2009).

———. "US Trains Colombian Military to Resist Rebels." *Austin American Statesman* (6 December 1998). http://proquest.umi.com/pqdweb?index=

0&did=36560244&SrchMode=2&sid=1&Fmt=3&VInst=PROD&VTy
pe=PQD&RQT=309&VName=PQD&TS=1242153130&clientId=5904
(accessed 5 November 2008).

Barber, Ben. "Colombia Needs Elections, Arms Aid, Ex-Army Chief Says."
Washington Times, 22 March 1998. http://proquest.umi.com/pqdw
eb?index=209&did=27643577&SrchMode=1&sid=1&Fmt=3&VI
nst=PROD&VType=PQD&RQT=309&VName=PQD&TS=12398
18287&clientId=5904 (accessed 5 November 2008).

Becker, David C. "Morphing War: Counter-narcotics, Counter-insurgency, and
Counter-terrorism Doctrine in Colombia." In *JSOU Report 1: 2004–
2005 JSOU/NDIA Essays.* Hulburt Field, FL: Joint Special Operations
University, 2005.

Bibes, Patricia. "Colombia: The Military and the Narco-Conflict." *Low Intensity
Conflict & Law Enforcement* (Spring 2000): 32–48.

Boraz, Steven C. "Intelligence Reform in Colombia: Transparency and
Effectiveness against Internal Threats." *Strategic Insights* (May 2007).

Bowden, Mark. *Killing Pablo: The Hunt for the World's Greatest Outlaw.* New
York, NY: Penguin Books, 2002.

Briscoe, Charles H. "Across the Pacific to War: The Colombian Navy in Korea,
1951–1955." *Veritas, Journal of Army Special Operations History* 2, no.
4 (2006): 24–29.

———. "Barbula and Old Baldy, March 1953: Colombia's Heaviest Combat in
Korea." *Veritas, Journal of Army Special Operations History* 2, no. 4
(2006): 15–23.

———. "Blue Helmets to Maroon Berets: *Batallón Colombia* in the Suez and
Sinai, 1956–1958, 1982–2006." *Veritas, Journal of Army Special
Operations History* 2, no. 4 (2006): 103–106

———. "Colombian *Lancero* School Roots." *Veritas, Journal of Army Special
Operations History* 2, no. 4 (2006): 30–37.

———. "*Plan Lazo*: Evaluation and Execution." *Veritas, Journal of Army Special
Operations History* 2, no. 4 (2006): 38–46.

Brodzinsky, Sibylla. "Colombia Turns Down Dilapidated US Trucks: Repairs
Too Costly Despite 'Donation.'" *Washington Times,* 2 December 1999.
http://proquest.umi.com/pqdweb?index=1&did=46807467&SrchMo
de=2&sid=1&Fmt=3&VInst=PROD&VType=PQD&RQT=309&VN
ame=PQD&TS=1242421733&clientId=5904 (accessed 5 November
1999).

Bronstein, Hugh, Joshua Goodman, and Bill Faries. "Top Rebel Commander Killed
by His Own Bodyguard, Colombia Says; Second FARC Leader Killed as
Countries Agree to Ease Tensions." *Ottawa Citizen*, 8 March 2008. http://
proquest.umi.com/pqdweb?index=7&did=1443445081&SrchMode=2&
sid=4&Fmt=3&VInst=PROD&VType=PQD&RQT=309&VName=PQ
D&TS=1247498664&clientId=5904 (accessed 3 November 2008).

Brooks, Laura. "Colombia Disbands Controversial Army Brigade." *Washington
Post,* 21 May 1998. http://proquest.umi.com/pqdweb?index=119&di

d=29617803&SrchMode=1&sid=1&Fmt=3&VInst=PROD&VType =PQD&RQT=309&VName=PQD&TS=1239820613&clientId=5904 (accessed 4 November 2008).

———. "Colombian Military Is Called to Account: Rebels Outwitted Forces, Critics Say." *Washington Post*, 8 March 1998. http://proquest.umi.com/ pqdweb?index=5&did=27000487&SrchMode=2&sid=3&Fmt=3&VI nst=PROD&VType=PQD&RQT=309&VName=PQD&TS=1241191 019&clientId=5904 (accessed 5 November 2008).

———. "Debacle in Colombia Shows Rebel Strength." *Washington Post,* 6 August 1998. http://proquest.umi.com/pqdweb?index=18&did=3266 1376&SrchMode=2&sid=17&Fmt=3&VInst=PROD&VType=PQD& RQT=309&VName=PQD&TS=1236966447&clientId=5904 (accessed 5 November 2008).

Brown, Tom. "Colombia Official Reveals Shake-up in Military's Echelon." *Houston Chronicle*, 1 November 1996. http://proquest.umi.com/pqd web?index=0&did=23117318&SrchMode=2&sid=1&Fmt=3&VIns t=PROD&VType=PQD&RQT=309&VName=PQD&TS=1240931 257&clientId=5904 (accessed 4 November 2008).

Bruneau, Thomas C. "Restructuring Colombia's Defense Establishment to Improve Civilian Control and Military Effectiveness." Monterey, CA: Center for Civil-Military Relations, Naval Postgraduate School, 2004. http://www. resdal.org/experiencias/main-lasa-04.html (accessed 11 March 2009).

Burns, Jimmy, and Adam Thomson. "Shades of Vietnam in Anti-Drug Fights: US Army Officers Are Joining Colombia's Action Against Narcotics—But It Will Involve Them in Conflict with Rebel Groups Too." *Financial Times,* 26 October 1999. http://lumen.cgsccarl.com/login?url=http:// proquest.umi/com/pqdweb?did=45794559&sid=2&Fmt=3&clientld=50 94&RQT=309&VName=PQD (accessed 5 November 2009).

Center for International Policy's Colombia Program. "Below the Radar: US Military Programs with Latin America, 1997–2007." Center for International Policy, Latin American Working Group Education Fund, and Washington Office on Latin America, March 2007.

Central Intelligence Agency. "The World Factbook: Colombia." https://www. cia.gov/library/publications/the-world-factbook/geos/co.html (accessed 7 August 2008).

Coleman, Bradley L. *Colombia and the United States: The Making of an Inter-American Alliance, 1939–1960*. Kent, OH: The Kent State University Press, 2008.

———. "The Colombian Army in Korea, 1950–1954." *The Journal of Military History* (January 2005): 1137–1177.

"Colombia Appoints New Head of Armed Forces." *Xinhua News Agency*, 16 August 2006. http://proquest.umi.com/pqdweb?index=1&did=1095 412221&SrchMode=2&sid=9&Fmt=3&VInst=PROD&VType=PQD& RQT=309&VName=PQD&TS=1247083356&clientId=5904 (accessed 8 July 2009).

"Colombia: Armed Forces Carry Out 'Unprecedented' Anti-Drug Operation in South." *BBC Monitoring Americas—Political,* 13 July 2000. http://proquest.umi.com/pqdweb?index=1&did=56418519&SrchMode=2&sid= 1&Fmt=3&VInst=PROD&VType=PQD&RQT=309&VName=PQD&TS=1242830120&clientId=5904 (accessed 4 November 2008).

"Colombia: Armed Forces Restructured." Stratfor Global Intelligence, 5 May 2005. http://www.stratfor.com/memberships/63192/colombia_armed_forces_restructured (accessed 2 July 2009).

"Colombia: Army Commander Says 13,000 Troops Advancing Toward Demilitarized Zone." *BBC Monitoring Americas—Political,* 14 January 2002. http://proquest.umi.com/pqdweb?index=1&did= 99540960&SrchMode=2&sid =1&Fmt=3&VInst=PROD&VType=PQD&RQT=309&VName=PQD&TS=1243527326&clientId=5904 (accessed 4 November 2008).

"Colombia Buys 25 Super Tucanos." *Defesa@Net,* 8 December 2005. http://www.defesanet.com.br/embraer/super_tucano_colombia7_e.htm (accessed 11 March 2009).

"Colombia: Commander Accepts Responsibility for Errors, Army's Reorganization." *BBC Monitoring Americas—Political,* 15 November 1998. "Colombian Soldiers Re-enter Disputed City/Rebels' Attack Raise Doubts About Peace Bid." http://proquest.umi.com/pqdweb?index=0&did=35964886&SrchMode=2&sid=2&Fmt=3&VInst=PROD&VType=PQD&RQT=309&VName=PQD&TS=1242143016&clientId=5904 (accessed 5 November 2008).

"Colombia: Commander Says Military to Ask for More Money." *BBC Monitoring Americas—Political,* 26 March 2002. http://proquest.umi.com/pqdweb?index=0&did=111527883&SrchMode=2&sid=3&Fmt=3&VInst=PROD&VType=PQD&RQT=309&VName=PQD&TS=1243536700&clientId=5904 (accessed 4 November 2008).

"Colombia: Further Details of FARC Attacks on Vigia del Fuerte and Elsewhere." *BBC Monitoring Americas—Political,* 28 March 2000. http://proquest.umi.com/pqdweb?RQT=403&TS=1237824812&clientId=5904&DBId=G5&xsq=Colombia&xFO=CITABS&xsq1=military&xFO1=CITABS&xOP1=AND&saved=1 (accessed 19 May 2009).

"Colombia: General Reacts to Public Reprimand of Armed Forces by President Uribe." *BBC Monitoring Americas—Political,* 29 August 2002. http://proquest.umi.com/pqdweb?index=0&did=156290021&SrchMode=2&sid=1&Fmt=3&VInst=PROD&VType=PQD&RQT=309&VName=PQD&TS=1245779069&clientId=5904 (accessed 4 November 2008).

"Colombia: Government Dismisses 388 Officers, NCOs." *BBC Monitoring Americas—Political,* 17 October 2000. http://proquest.umi.com/pqdweb?index=1&did=62614800&SrchMode=2&sid=1&Fmt=3&VInst=PROD&VType=PQD&RQT=309&VName=PQD&TS=1242839474&clientId=5904 (accessed 4 November 2008).

"Colombia: Investigations Show FARC Controlled Drug Crops, Trafficking in Ex-DMZ." *BBC Monitoring Americas—Political,* 15 March 2002. http://

proquest.umi.com/pqdweb?index=0&did=110486053&SrchMode=2& sid=8&Fmt=3&VInst=PROD&VType=PQD&RQT=309&VName=PQ D&TS=1243542032&clientId=5904 (accessed 4 November 2008).

"Colombia: Joint Operation Against FARC's Teofilo Forero Unit Bags 52 Guerrillas." *BBC Monitoring Americas—Political*, 3 June 2003. http:// proquest.umi.com/pqdweb?index=0&did=343349231&SrchMode=2& sid=3&Fmt=3&VInst=PROD&VType=PQD&RQT=309&VName=PQ D&TS=1246115960&clientId=5904 (accessed 4 November 2008).

"Colombian Military Chief Says Armed Forces Close to Defeating Guerrillas." *BBC Monitoring Americas*, 17 September 2007. http://proquest.umi. com/pqdweb?index=0&did=1337048011&SrchMode=2&sid=1&Fmt= 3&VInst=PROD&VType=PQD&RQT=309&VName=PQD&TS=1247 330956&clientId=5904 (accessed 3 November 2008).

"Colombia: Military Forces Chief Asks Generals to Locate, Rescue All Hostages." *BBC Monitoring Americas*, 27 November 2003. http://proquest.umi. com/pqdweb?index=0&did=465970841&SrchMode=2&sid=5&Fmt= 3&VInst=PROD&VType=PQD&RQT=309&VName=PQD&TS=1246 372576&clientId=5904 (accessed 4 November 2008).

"Colombia: Military 'Summit' Analyses Tasks; Four Mobile Brigades Being Set Up." *BBC Monitoring Americas—Political*, 27 January 2001. http:// proquest.umi.com/pqdweb?index=0&did=67351439&SrchMode=2&si d=2&Fmt=3&VInst=PROD&VType=PQD&RQT=309&VName=PQD &TS=1243027271&clientId=5904 (accessed 4 November 2008).

"Colombia Names New Armed Forces Leader." *New York Times*, 19 November 2003. http://proquest.umi.com/pqdweb?index=1&did=453199511&Src hMode=2&sid=2&Fmt=3&VInst=PROD&V.Type=PQD&RQT=309& VName=PQD&TS=1246371594&clientId=5904 (accessed 4 November 2008).

"Colombia: New Commanders of Divisions Attend Military 'Summit.'" *BBC Monitoring Americas—Political*, 18 January 2001. http://proquest.umi. com/pqdweb?index=2&did=66907118&SrchMode=2&sid=1&Fmt= 3&VInst=PROD&VType=PQD&RQT=309&VName=PQD&TS=1243 026063&clientId=5904 (accessed 4 November 2008).

"Colombia: Pastrana Praises Armed Forces, Police for their Anti-Rebel Efforts." *BBC Monitoring Americas—Political*, 22 July 2002. http://proquest.umi. com/pqdweb?index=0&did=140519251&SrchMode=2&sid=1&Fmt=3 &VInst=PROD&VType=PQD&RQT=309&VName=PQD&TS=12436 97101&clientId=5904 (accessed 4 November 2008).

"Colombia: Peasant Soldiers Return Home after Completing Military Training." *BBC Monitoring Americas—Political*, 6 March 2003. http://proquest. umi.com/pqdweb?index=0&did=301760851&SrchMode=2&sid=1&F mt=3&VInst=PROD&VType=PQD&RQT=309&VName=PQD&TS=1 246135556&clientId=5904 (accessed 4 November 2008).

"Colombia: President Pastrana's Handling of Military Crisis Assessed." *BBC Monitoring Americas—Political*, 14 December 1999. http://proquest.

umi.com/pqdweb?index=3&did=47218556&SrchMode=2&sid=1&Fmt
=3&VInst=PROD&VType=PQD&RQT=309&VName=PQD&TS=124
2419601&clientId=5904 (accessed 5 November 2008).

"Colombia: River Brigade Forces Begin Operations on 4th August." *BBC Monitoring Americas—Political,* 1 August 1999. http://proquest.umi.com/pqdweb?did=43586599&sid=1&Fmt=3&clientId=5094&RQT=309&VName=PQD (accessed 5 November 2008).

"Colombia Says Paramilitary Demobilization Complete," *BBC Monitoring Newsfile,* 18 April 2008, http://proquest.umi.com/pqdweb?index=0&did=1022593911&SrchMode=2&sid=1&Fmt=3&VInst=PROD&VType=PQD&RQT=309&VName=PQD&TS=1247072929&clientId=5904 (accessed 4 November 2008).

"Colombia: Two Analysts See Serious Problems with Government's Security Policy." *BBC Monitoring Americas*, 20 April 2004. http://proquest.umi.com/pqdweb?index=0&did=621176981&SrchMode=2&sid=1&Fmt=3&VInst=PROD&VType=PQD&RQT=309&VName=PQD&TS=1246553296&clientId=5904 (accessed 4 November 2008).

"Colombia: US Insists on Approval of Any Action to Free US Hostages," *BBC Monitoring Americas—Political*, 20 May 2003. http://proquest.umi.com/pqdweb?index=0&did=338532281&SrchMode=2&sid=5&Fmt=3&VInst=PROD&VType=PQD&RQT=309&VName=PQD&TS=1246116365&clientId=5904 (accessed 4 November 2008).

"Colombia Targets Rebel Stronghold: 5,000-Strong Unit Will Patrol Southern Jungle." *Rocky Mountain News,* 13 March 1998. http://proquest.umi.com/pqdweb?did=28564227&sid=3&Fmt=3&clientId=5094&RQT=309&VName=PQD (accessed 5 November 2008).

"Colombian Air Force Chief: Several Guerrilla Leaders Targeted." *BBC Monitoring Americas—Political,* 4 March 2002. http://proquest.umi.com/pqdweb?index=1&did=110023051&SrchMode=2&sid=4&Fmt=3&VInst=PROD&VType=PQD&RQT=309&VName=PQD&TS=1243540735&clientId=5904 (accessed 4 November 2008).

"Colombian Army Says 40 Killed in Clash With Rebels." *Washington Post,* 17 August 1998. http://proquest.umi.com/pqdweb?index=1&did=33165732&SrchMode=2&sid=1&Fmt=3&VInst=PROD&VType=PQD&RQT=309&VName=PQD&TS=1242064952&clientId=5904 (accessed 11 May 2009).

"Colombian Government Issues Official Report on 'FARC Genocide.'" *BBC Monitoring Newsfile,* 6 May 2003. http://proquest.umi.com/pqdweb?index=0&did=332739061&SrchMode=2&sid=1&Fmt=3&VInst=PROD&VType=PQD&RQT=309&VName=PQD&TS=1246114047&clientId=5904 (accessed 4 November 2008).

"Colombian Guerrillas Ambush Army Convoy, Kill 31 Soldiers." *San Antonio Express–News,* 17 April 1996. http://proquest.umi.com/pqdweb?index=8&did=15244774&SrchMode=2&sid=2&Fmt=3&VInst=PROD&V

Type=PQD&RQT=309&VName=PQD&TS=1241197937&clientId=
5904 (accessed 17 November 2008).

"Colombian Military Hold Emergency Meeting to Debate Counterterrorism Strategy." *BBB Monitoring Americas—Political,* 13 February 2002. http://proquest.umi.com/pqdweb?index=0&did=107302002&Srch Mode=2&sid=5&Fmt=3&VInst=PROD&VType=PQD&RQT=309&V Name=PQD&TS=1243525103&clientId=5904 (accessed 4 November 2008).

"Colombian Military Links to Executions Could Hinder Human Rights Certification." *BBC Monitoring Americas*, 25 June 2007. http://proquest. umi.com/pqdweb?index=0&did=1294313041&SrchMode=2&sid=3&F mt=3&VInst=PROD&VType=PQD&RQT=309&VName=PQD&TS= 1247325082&clientId=5904 (accessed 3 November 2008).

"Colombian Military: Pastrana Counterterrorism Bid Bound to Fail." *BBC Monitoring Americas—Political,* 6 May 2002. http://proquest.umi.com/ pqdweb?index=0&did=117885474&SrchMode=2&sid=2&Fmt=3& VInst=PROD&VType=PQD&RQT=309&VName=PQD&TS=124362 6112&clientId=5904 (accessed 4 November 2008).

"Colombian Military Plan Yet to Defeat FARC—Political Analyst." *BBC Monitoring Americas*, 20 January 2008. http://proquest.umi.com/pqd web?did=1415609471&sid=3&Fmt=3&clientId=5094&RQT=309& VName=PQD (accessed 3 November 2008).

"Colombian Minister Stresses Importance of Role of State in Security Issues." *BBC Monitoring Americas—Political*, 21 October 2002. http://proquest. umi.com/pqdweb?index=19&did=217198381&SrchMode=2&sid=15& Fmt=3&VInst=PROD&VType=PQD&RQT=309&VName=PQD&TS= 1236966153&clientId=5904 (accessed 4 November 2008).

"Colombian Public Losing Confidence in Military Intelligence, Daily Reports." *BBC Monitoring Americas*, 27 September 2006. http://proquest.umi. com/pqdweb?index=0&did=1136001071&SrchMode=2&sid=1&Fmt= 3&VInst=PROD&VType=PQD&RQT=309&VName=PQD&TS=1247 155165&clientId=5904 (accessed 4 November 2008).

"Colombian Soldiers Re-enter Disputed City/Rebels' Attack Raise Doubts About Peace Bid." *Houston Chronicle,* 5 November 1998. http://proquest.umi. com/pqdweb?index=8&did=35718239&SrchMode=2&sid=1&Fmt= 3&VInst=PROD&VType=PQD&RQT=309&VName=PQD&TS=1242 141204&clientId=5904 (accessed 5 May 2009).

"Colombia's Brass Admit Army Cannot Compete with Guerrillas." *Houston Chronicle*, 25 March 1998. http://proquest.umi.com/pqdweb?index= 1&did=27791616&SrchMode=2&sid=1&Fmt=3&VInst=PROD&VTy pe=PQD&RQT=309&VName=PQD&TS=1240519282&clientId=5904 (accessed 5 November 2008).

"Colombia's Defense Minister Quits over Concession to Rebels." *New York Times,* 27 May 1999. http://proquest.umi.com/pqdweb?index=13&did=4191

2447&SrchMode=2&sid=3&Fmt=3&VInst=PROD&VType=PQD&R
QT=309&VName=PQD&TS=1242327099&clientId=5904   (accessed
14 May 2009).

"Colombia's FARC Seen as Considerably Weakened But Not Defeated." *BBC Monitoring Americas*, 4 December 2007. http://proquest.umi.com/pqdw eb?index=1&did=1392730681&SrchMode=2&sid=1&Fmt=3&VInst= PROD&VType=PQD&RQT=309&VName=PQD&TS=1247343082&c lientId=5904 (accessed 3 November 2008).

"Colombia's New Armed Forces Chief 'Strategic' Intelligence Expert." *BBC Monitoring Americas*, 22 August 2006. http://proquest.umi.com/pqdw eb?index=0&did=1102991381&SrchMode=2&sid=1&Fmt=3&VInst= PROD&VType=PQD&RQT=309&VName=PQD&TS=1247680055&c lientId=5904 (accessed 4 November 2008).

"Colombia's Pastrana Beefs Up Military, Offers Reward for Rebel Leaders' Arrest." *BBC Monitoring Americas—Political,* 29 June 2002. http:// proquest.umi.com/pqdweb?index=0&did=130362711&SrchMode=2& sid=1&Fmt=3&VInst=PROD&VType=PQD&RQT=309&VName=PQ D&TS=1243693667&clientId=5904 (accessed 4 November 2008).

"Commander Defends Colombian Armed Forces in Interview." *BBC Monitoring Americas—Political,* 10 November 2006. http://proquest.umi.com/ pqdweb?index=0&did=1159878151&SrchMode=2&sid=21&Fmt=3 &VInst=PROD&VType=PQD&RQT=309&VName=PQD&TS=1236 966958&clientId=5904 (accessed 4 November 2008).

Cook, Dean A. "U.S. Southern Command: General Charles E. Wilhelm and the Shaping of U.S. Military Engagement in Colombia, 1997–2000." In *America's Viceroys: The Military and U.S. Foreign Policy*, edited by Derek S. Reveron, 127–162. New York, NY: Palgrave Macmillan, 2004.

Cope, John A. "Colombia's War: Toward a New Strategy." Washington, DC: National Defense University, Institute for National Strategic Studies, Strategic Forum 194 (October 2002).

Darling, Juanita, and Ruth Morris. "Crash Points to Military Role of US in Colombia." *Los Angeles Times,* 28 July 1999. http://proquest.umi.com/ pqdweb?index=17&did=43516995&SrchMode=2&sid=4&Fmt=3&V Inst=PROD&VType=PQD&RQT=309&VName=PQD&TS=124241 3974&clientId=5904 (accessed 5 November 2008).

de Corba, Jose. "Guerrilla Attacks Kill 100 in Colombia in Possible Link to Nation's Drug War." *Wall Street Journal,* 3 September 1996. http:// proquest.umi.com/pqdweb?index=10&did=23868034&SrchMode=2& sid=2&Fmt=3&VInst=PROD&VType=PQD&RQT=309&VName=PQ D&TS=1241204004&clientId=5904 (accessed 5 November 2008).

de Cordoba, Jose. "Southern Front: Rebels Flail in Colombia after Death of Leader." *Wall Street Journal*, 28 May 2008. http://proquest.umi.com/ pqdweb?did=1485594701&sid=1&Fmt=4&clientld-5094&RQT= 309&VName=PQD (accessed 3 November 2008).

"Defense Minister Says Security Forces Must be Present Throughout Colombia."
BBC Monitoring Americas—Political, 26 August 2002. http://proquest.
umi.com/pqdweb?index=0&did=155320491&SrchMode=2&sid=1&
Fmt=3&VInst=PROD&VType=PQD&RQT=309&VName=PQD&TS=
1245775397&clientId=5904 (accessed 4 November 2008).

de la Garza, Paul, and David Adams, "Military Aid . . . From the Public Sector."
St. Petersburg Times, 3 December 2000. http://proquest.umi.com/pqd
web?index=3&did=64917780&SrchMode=2&sid=1&Fmt=3&VInst=
PROD&VType=PQD&RQT=309&VName=PQD&TS=1243349412&c
lientId=5904 (accessed 4 March 2009).

Demarest, Geoffrey B. "The Overlap of Military and Police in Latin America," April
1995. http://www.smallwars.quantico.usmc.mil/search/LessonsLearned/
LatinAm/milpolre.htm (accessed 29 August 2008).

"Democratic Security Policy to Continue, Says Colombia's New Defence
Minister." *BBC Monitoring Americas*, 26 July 2006. http://proquest.umi.
com/pqdweb?index=0&did=1083390701&SrchMode=2&sid=7&Fmt=
3&VInst=PROD&VType=PQD&RQT=309&VName=PQD&TS=1247
081947&clientId=5904 (assessed 8 July 2009).

DeShazo, Peter, Phillip McLean, and Johanna Mendelson Formal. *Colombia's Plan
de Consolidacion Integral de la Macarena: An Assessment.* Washington,
DC: Center for Strategic and International Studies, June 2009.

DeShazo, Peter, Tanya Primiani, and Philip McLean. *Back From the Brink:
Evaluating Colombia, 1999–2007.* Washington, DC: Center for Strategic
and International Studies, November 2007.

Downes, Richard. *Landpower and Ambiguous Warfare: The Challenge of
Colombia in the 21st Century—Conference Report.* Carlisle Barracks,
PA: Strategic Studies Institute, 10 March 1999.

Dudley, Steven. "Colombian President Alienates Military Leaders." *McClatchey
Tribune News Service*, 21 March 2006. http://proquest.umi.com/pqd
web?index=1&did=1007298581&SrchMode=2&sid=1&Fmt=3&V
Inst=PROD&VType=PQD&RQT=309&VName=PQD&TS=12469
99653&clientId=5904 (accessed 4 November 2008).

Escuela de Infanteria del Ejercito. *Libro de la Infanteria.* Bogotá: Escuela de
Infanteria del Ejercito, 2008.

Evans, Michael, ed. "War in Colombia: Guerrillas, Drugs and Human Rights
in US-Colombia Policy, 1988–2002." National Security Archive
Electronic Briefing Book No. 69, Colombia Documentation Project.
http://www.gwu.edu/~nsarchiv/NSAEBB/NSAEBB69/ (accessed 4
August 2008).

Farah, Douglas. "Colombian Army Fighting Legacy of Abuses." *Washington
Post,* 18 February 1999. http://proquest.umi.com/pqdweb?index=2&d
id=39087527&SrchMode=2&sid=1&Fmt=3&VInst=PROD&VType
=PQD&RQT=309&VName=PQD&TS=1242237748&clientId=5904
(accessed 5 November 2008).

————. "Colombian Rebels Seen Winning War; US Study Finds Army Inept, Ill-Equipped." *Washington Post,* 10 April 1998. http://proquest.umi.com/pqdweb?index=9&did=28551847&SrchMode=2&sid=2&Fmt=3&VInst=PROD&VType=PQD&RQT=309&VName=PQD&TS=1237824894&clientId=5904 (accessed 5 November 2008).

————. "US to Aid Colombian Military: Drug-Dealing Rebels Take Toll on Army." *Washington Post,* 27 December 1998. http://proquest.umi.com/pqdweb?index=5&did=37803016&SrchMode=2&sid=22&Fmt=3&VInst=PROD&VType=PQD&RQT=309&VName=PQD&TS=1236967126&clientId=5904 (accessed 5 November 2008).

"FARC Attacks Seek to Discredit President, Colombian Weekly Says." *BBC Monitoring Americas*, 4 July 2005. http://proquest.umi.com/pqdweb?index=0&did=862530111&SrchMode=2&sid=1&Fmt=3&VInst=PROD&VType=PQD&RQT=309&VName=PQD&TS=1246639073&clientId=5904 (accessed 4 November 2008).

"FARC Offensive Signals Change in Tactics, Colombian Weekly Says." *BBC Monitoring Americas*, 14 February 2005. http://proquest.umi.com/pqdweb?index=0&did=792945081&SrchMode=2&sid=3&Fmt=3&VInst=PROD&VType=PQD&RQT=309&VName=PQD&TS=1246639601&clientId=5904 (accessed 4 November 2008).

Farrell, Michelle L. "Sequencing: Targeting Insurgents and Drugs in Colombia." Monterey, CA: Naval Post Graduate School Thesis, March 2007.

Feldmann, Andreas E., and Victor J. Hinojosa. "Terrorism in Colombia: Logic and Sources of a Multidimensional and Ubiquitous Phenomenon." *Terrorism and Political Violence* 21 (2009): 42–61.

Ferrer, Yadira. "Change of Commander in Chief a Boost to Peace." *Inter Press Service* (25 July 1997). http://proquest.umi.com/pqdweb?did=13182995&sid=4&Fmt=3&clientId=5094&RQT=309&VName=PQD (accessed 5 November 2008).

————. "Colombia: Uribe Launches Controversial Network of Informers." *Global Information Network* (9 August 2002). http://proquest.umi.com/pqdweb?did=321047661&sid=1&Fmt=3&clientId=5094&RQT=309&VName=PQD (accessed 4 November 2008).

Finlayson, Kenneth. "Colombia: A Special Relationship." *Veritas, Journal of Army Special Operations History* 2, no. 4 (2006): 5–7.

————. "Colombian Special Operations Forces." *Veritas, Journal of Army Special Operations History* 2, no. 4 (2006): 56–59.

————. "'Conducting the Orchestra:' AOB 740 in Colombia." *Veritas, Journal of Army Special Operations History* 2, no. 4 (2006): 69–73.

————. "OPATT to PATT: El Salvador to Colombian and the Formation of Planning and Assistance Training Teams." *Veritas, Journal of Army Special Operations History* 2, no. 4 (2006): 91–94.

————. "'There Is a Word I Need to Learn:' ODA 741 and Colombian National Police Training at Espinal," *Veritas, Journal of Army Special Operations History* 2, no. 4 (2006): 79–84.

Fishel, John T. "Colombia: Civil-Military Relations in the Midst of War." *Joint Force Quarterly* (Summer 2000): 51–56.

Fleming, Adam L. "Colombia's Resurrection: Alternative Development is the Key to Democratic Security." Monterey, CA: Naval Postgraduate School Thesis, September 2004.

Flores, Ricardo A. "Improving the US Navy Riverine Capability: Lessons from the Colombian Experience." Monterey, CA: Naval Postgraduate School Thesis, December 2007.

Forero, Juan. "Burdened Colombians Back Tax to Fight Rebels." *New York Times,* 8 September 2002. http://proquest.umi.com/pqdweb?index=1&did=163 053271&SrchMode=2&sid=2&Fmt=3&VInst=PROD&VType=PQD& RQT=309&VName=PQD&TS=1245780557&clientId=5904 (accessed 4 November 2008).

———. "Colombia Fires 27 From Army Over Killings; Youths' Deaths Attributed to Stress on Body Counts." *Washington Post*, 30 October 2008. http:// proquest.umi.com/pqdweb?index=7&did=1585149411&sod=6&Fmt =3&clientId=5094&RQT=309&VName=PQD (accessed 3 November 2008).

———. "Colombia Says Rebels Have Killed 56 Troops." *New York Times,* 21 October 2000. http://proquest.umi.com/pqdweb?index=39&did=62 797057&SrchMode=1&sid=4&Fmt=3&VInst=PROD&VType=PQD& RQT=309&VName=PQD&TS=1242938270&clientId=5904 (accessed 5 November 2008).

———. "Colombia's Army Rebuilds and Challenges Rebels." *New York Times,* 2 September 2001. http://proquest.umi.com/pqdweb?index=0&did=79 411237&SrchMode=2&sid=1&Fmt=3&VInst=PROD&VType=PQD& RQT=309&VName=PQD&TS=1243439417&clientId=5904 (accessed 4 November 2008).

———. "Colombia Tries Social Reform Programs As New Weapons Against Rebels; Aim Is to Establish a State Presence in Long-Neglected Areas." *Washington Post*, 10 July 2007. http://proquest.umi.com/pqdweb?index= 0&did=1301875571&SrchMode=2&sid=1&Fmt=3&VInst=PROD& VType=PQD&RQT=309&VName=PQD&TS=1247333209&client Id=5904 (accessed 3 November 2008).

———. "Colombian Army Goes High Up to Fight Rebels." *New York Times,* 19 December 2000. http://proquest.umi.com/pqdweb?index=0&did=65 287103&SrchMode=2&sid=1&Fmt=3&VInst=PROD&VType=PQD& RQT=309&VName=PQD&TS=1242942108&clientId=5904 (accessed 4 November 2008).

———. "Colombian Leader Says Rebels Killed 11 Civilian Hostages." *Washington Post*, 29 June 2007. http://proquest.umi.com/pqdweb?index=7&did= 1296614111&SrchMode=2&sid=1&Fmt=3&VInst=PROD&VType= PQD&RQT=309&VName=PQD&TS=1247327973&clientId=5904 (accessed 3 November 2008).

─────. "Colombian Rebel Commander Killed; Strike by Government Forces Called Major Setback for FARC Guerrillas." *Washington Post*, 2 March 2008. http://proquest.umi.com/pqdweb?index=17&did=1438193001 &SrchMode=2&sid=1&Fmt=3&VInst=PROD&VType=PQD&RQT =309&VName=PQD&TS=1247497351&clientId=5904 (accessed 3 November 2008).

─────. "Colombian Troops Kill Farmers, Pass Off Bodies as Rebels.'" *Washington Post*, 30 March 2008. http://proquest.umi.com/pqdweb?index=2&did =1453809271&SrchMode=2&sid=1&Fmt=3&VInst=PROD&VType =PQD&RQT=309&VName=PQD&TS=1247514403&clientId=5904 (accessed 3 November 2008).

─────. "Colombian Troops Move on Rebel Zone as Talks Fail." *New York Times*, 11 January 2002. http://proquest.umi.com/pqdweb?index=3&did=990 67316&SrchMode=2&sid=6&Fmt=3&VInst=PROD&VType=PQD& RQT=309&VName=PQD&TS=1243525371&clientId=5904 (accessed 4 November 2008).

─────. "In Colombia Jungle Ruse, US Played a Quite Role; Ambassador Spotlights Years of Aid, Training." *Washington Post*, 9 July 2008. http:// proquest.umi.com/pqdweb?index=3&did=1507416311&SrchMode= 2&sid=3&Fmt=3&VInst=PROD&VType=PQD&RQT=309&VName= PQD&TS=1247586488&clientId=5904 (accessed 3 November 2008).

─────. "Rebel Force in Colombia Repatriates 242 More P.O.W.'s." *New York Times*, 29 June 2001. http://proquest.umi.com/pqdweb?index=2&did=74 837004&SrchMode=2&sid=1&Fmt=3&VInst=PROD&VType=PQD& RQT=309&VName=PQD&TS=1243371882&clientId=5904 (accessed 4 November 2008).

Franco, George. "Implementing Plan Colombia: Assessing the Security Forces Campaign." *Special Warfare* (Winter 2002): 28–35.

Frechette, Myles R.R. *Colombia and the United States—The Partnership: But What is the Endgame?* Carlisle Barracks, PA: Strategic Studies Institute, February 2007.

Gamini, Gabriella. "Colombia Fights Back as Guerrilla Launch Fiercest Raids in Decades." *Times*, 3 September 1996. http://proquest.umi.com/pqdweb? did=34285574&sid=5&Fmt=3&clientId=5094&RQT=309&VName= PQD (accessed 5 November 2008).

Gates, Robert M. "US Global Leadership Campaign." Speech delivered 15 July 2008. http://www.defenselink. mil/utility/printitem.aspx?print=http:// www.defenselink.mil/speeches/speech.aspx?speechid=1262 (accessed 15 July 2009).

"General Montoya's Replacement." *Semana.com*, 10 November 2008, http:// www.semana.com/wf_ImprimirArticuloIngles.aspx?IdArt=117595 (accessed 15 January 2009).

Geyer, Georgie. "For Colombia, 1996 Nothing But Trouble." *Tulsa World*, 31 January 1997. http://proquest.umi.com/pqdweb?index=192&did=1682 4144&SrchMode=1&sid=7&Fmt=3&VInst=PROD&VType=PQD&R

QT=309&VName=PQD&TS=1239811252&clientld=5904   (accessed 5 November 2008).

Hanratty, Dennis M., and Sandra W. Meditz, ed. *Colombia: A Country Study.* Washington, DC: Federal Research Division of the Library of Congress, December 1988.

Jane's. "Fuerzas Armadas Revolucionarias de Colombia (FARC)." http://www8. janes.com/Search/documentView. do?docId=/content1/janesdata/binder/ jwit/jwit0265.htm@current&pageSelected=allJanes&backPath=http: //search.janes.com/Search&Prod_Name=JWIT&keyword   (accessed 20 November 2008).

Jane's. "Jane's Sentinel Security Assessment—South America: Colombia." http://www4.janes.com/subscribe/sentinel/SAMS_doc_view.jsp? Sent_Country=Colombia&Prod_Name=SAMS&K2DocKey=/ content1/janesdata/sent/samsu/colos010.htm@current#toclink-j1511111681950216 (accessed 20 November 2008).

Jane's Sentinel Country Risk Assessment. http://sentinel.janes.com/docs/sentinel/ SAMS_country.jsp?Prod_Name=SAMS&Sent_Country=Colombia& (accessed 20 November 2008).

Jenkins, Brian M. "Colombia: Crossing a Dangerous Threshold." *The National Interest* (Winter 2000/2001): 47–55.

Johnson, Stephen. "Helping Colombia Fix Its Plan to Curb Drug Trafficking, Violence, and Insurgency." Washington, DC: The Heritage Foundation Backgrounder, No. 1435, 26 April 2001. http://www.heritage.org/ Research/LatinAmerica/bg1887.cfm (accessed 15 September 2008).

Jones, Robert W. Jr. "Special Forces in Larandia: ODA 753 and the CERTE." *Veritas, Journal of Army Special Operations History* 2, no. 4 (2006): 85–90.

———. "'Who Taught Those Guys to Shoot Like Chuck Norris?' ODA 746 in Tolemaida." *Veritas, Journal of Army Special Operations History* 2, no. 4 (2006): 73–78.

Kaplan, Robert D. *Imperial Grunts: The American Military on the Ground.* New York, NY: Random House, 2005.

Kotler, Jared. "At Least 70 Dead in Colombia Rebel Attack: Assault on the Police Station Was Waged Even as Peace Talks Approach." *Fresno Bee,* 3 November 1998. http://proquest.umi.com/pqdweb?index=2&did=356 84454&SrchMode=2&sid=3&Fmt=3&VInst=PROD&VType=PQD& RQT=309&VName=PQD&TS=1242147398&clientId=5904 (accessed 11 May 2009).

Kovaleski, Serge F. "Colombian Army Drill: Respect for Rights: Training Course Tests Empathy, Discipline." *Washington Post,* 29 August 1999. http:// proquest.umi.com/pqdweb?index=9&did=44275664&SrchMode=2& sid=1&Fmt=3&VInst=PROD&VType=PQD&RQT=309&VName=PQ D&TS=1242403982&clientId=5904 (accessed 5 November 2008).

Krauss, Clifford. "Attacks by Colombian Rebels Appear a Response to US Plan." *New York Times,* 20 July 2000. http://proquest.umi.com/pqdweb?index =0&did=56669258&SrchMode=2&sid=2&Fmt=3&VInst=PROD&V

Type=PQD&RQT=309&VName=PQD&TS=1242831399&clientId=5
904 (accessed 19 May 2009).

Lakshmanan, Indira A.R. "Rebels Reassert Deadly Agenda in Colombia." *Boston Globe*, 26 May 2005. http://proquest.umi.com/pqdweb?index=2&did =844886731&SrchMode=2&sid=5&Fmt=3&VInst=PROD&VType =PQD&RQT=309&VName=PQD&TS=1246640478&clientId=5904 (accessed 4 November 2008).

Leech, Garry. "US/COLOMBIA: Demobilizing the AUC?" *NACLA Report on the Americas*, September/October 2004. http://proquest.umi.com/ pqdweb?did=693199551&sid=2&Fmt=3&clientId=5094&RQT= 309&VName=PQD (accessed 4 November 2008).

Manwaring, Max G. *Non-State Actors in Colombia: Threats to the State and to the Hemisphere*. Carlisle Barracks, PA: Strategic Studies Institute, May 2002.

———. "United States Security Policy in the Western Hemisphere: Why Colombia, Why Now, and What is to be Done?" *Small Wars and Insurgencies* (Autumn 2001): 67–96.

Marcella, Gabriel. *The United States and Colombia: The Journey from Ambiguity to Strategic Clarity*. Carlisle Barracks, PA: Strategic Studies Institute, May 2003.

Marks, Thomas A. "A Model Counterinsurgency: Uribe's Colombia (2002–2006) vs FARC." *Military Review* (March–April 2007): 41–56.

———. *Colombian Army Adaptation to FARC Insurgency*. Carlisle Barracks, PA: Strategic Studies Institute, January 2002.

———. "Colombian Army Counterinsurgency." *Crime, Law & Social Change* (July 2003): 77–105.

———. "Colombian Military Support for 'Democratic Security.'" *Small Wars and Insurgencies* (June 2006): 197–220.

———. *Sustainability of Colombian Military/Strategic Support for "Democratic Security."* Carlisle Barracks, PA: Strategic Studies Institute, July 2005.

Martinez, Margarita. "Colombian Chief Takes Blame in Rescue Fiasco, Sound of Helicopters Set Off an Orgy of Death in the Jungle." *San Antonio Express–News*, 7 May 2003. http://proquest.umi.com/ pqdweb?index=3&did= 780326551&SrchMode=2&sid=2&Fmt=3&VInst=PROD&VType= PQD&RQT=309&VName=PQD&TS=1246114784&clientId=5904 (accessed 4 November 2008).

McLean, Phillip. "Focus on Latin America: Colombia is Complicated." *Foreign Service Journal* (April 2006): 33–38.

Maullin, Richard L. *Soldiers, Guerrillas and Politics in Colombia*. Santa Monica, CA: Rand, December 1971.

Miller, T. Christian. "THE WORLD; Colombian Air Force Chief Quits; General Resigns Amid US Pressure and New Evidence Suggesting that Pilots Knowingly Fired on Civilians during a 1998 Bombing Raid." *Los Angeles Times*, 26 August 2003. http://proquest.umi.com/pqdweb?index=2&did= 388820971&SrchMode=2&sid=5&Fmt=3&VInst=PROD&VType=

186

PQD&RQT=309&VName=PQD&TS=1246045550&clientId=5904 (accessed 4 November 2008).

Morgenstein, Jonathan. "Consolidating Disarmament: Lessons from Colombia's Reintegration Program for Demobilized Paramilitaries." Washington, DC: United States Institute of Peace Special Report 217, November 2008. http://www.usip.org/pubs/specialreports/sr217.html (accessed 18 March 2008).

Muller, Christopher W. "USMILGP Colombia: Transforming Security Cooperation in the Global War on Terrorism." Monterey, CA: Naval Postgraduate School Thesis, December 2006.

Orellana, Manuel A. "How to Train an Army of Intelligence Analysts." Monterey, CA: Naval Postgraduate School Thesis, September 2005.

Otis, John. "How Colombian Army Hoodwinked the Rebels." *Houston Chronicle*, 4 July 2008. http://proquest.umi.com/pqdweb?index=4&did=15054203 41&SrchMode=2&sid=1&Fmt=3&VInst=PROD&VType=PQD&RQ T=309&VName=PQD&TS=1247584749&clientId=5904 (accessed 3 November 2008).

Passage, David. *The United States and Colombia: Untying the Gordian Knot.* Carlisle Barracks, PA: Strategic Studies Institute, March 2000.

Peceny, Mark, and Michael Durnan. "The FARC's Best Friend: US Antidrug Policies and the Deepening of Colombia's Civil War in the 1990's." *Latin American Politics and Society* (Summer 2006): 95–116.

Perdomo, Jose R. "Colombia's Democratic Security and Defense Policy in the Demobilization of the Paramilitaries." US Army War College Student Paper, 29 March 2007.

Perez, William F. "An Effective Strategy for Colombia: A Potential End to the Current Crisis." US Army War College Student Paper, 3 May 2004.

Pizano, Eduardo. *Plan Colombia: The View from the Presidential Palace.* Carlisle Barracks, PA: Strategic Studies Institute, May 2001.

Porch, Douglas. "Uribe's Second Mandate, the War, and the Implications for Civil-Military Relations in Colombia." Naval Postgraduate School, Center for Contemporary Conflict, Strategic Insights 2 (February 2006). http://www.ccc.nps.navy.mil/si/2006/Feb/porchFeb06.asp (accessed 19 March 2009).

Porch, Douglas, and Christopher W. Muller. "'Imperial Grunts' Revisited: The US Advisory Mission in Colombia." In *Military Advising and Assistance: From Mercenaries to Privatization, 1815–2007,* edited by Donald Stoker, 168–191. London: Routledge, 2008.

Priest, Dana. *The Mission: Waging War and Keeping Peace with America's Military.* New York, NY: W.W. Norton and Company, 2004.

———. "US Force Training Troops in Colombia; Exercise Anti-Drug Efforts." *Washington Post,* 25 May 1998. http://proquest.umi.com/pqdweb?inde x=5&did=29679864&SrchMode=2&sid=1&Fmt=3&VInst=PROD& VType=PQD&RQT=309&VName=PQD&TS=1241189705&clientId= 5904 (accessed 5 November 2008).

Rabasa, Angel, and Peter Chalk. *Colombian Labyrinth: The Synergy of Drugs and Insurgency and Its Implications for Regional Stability*. Santa Monica, CA: Rand, 2001.

Radu, Michael. "E-Notes Colombia: A Trip Report." Foreign Policy Research Institute, 1 June 2001. http://www.fpri.org/enotes/20010601.radu.colombiatrip.html (accessed 6 August 2008).

———. "E-Notes Colombia: Lucidity at Last." Foreign Policy Research Institute, 1 March 2002. http://www.fpri.org/enotes/20020301.radu.colombialucidityatlast.html (accessed 6 August 2008).

Rangel Suarez, Alfredo. *Colombia: Balance de Seguridad 2001–2004*. Fundacion Seguridad y Democracia Conflicto Armado, 28 December 2004. http://www.seguridadydemocracia.org/docConflicto.asp (accessed 18 March 2009).

———. *Fuerzas Militares para la guerra: La agenda pendiente de la reforma militar*. Bogotá: Ensayos de Seguridad Y Democracia, November 2003.

———. "Parasites and Predators: Guerrillas and the Insurrection Economy of Colombia." *Journal of International Affairs* (Spring 2000): 577–601.

———. *Sostenibilidad De La Seguridad Democratica*. Bogotá: Fundadcion Seguridad & Democracia, 23 February 2005.

Rempe, Dennis M. "Guerrillas, Bandits, and Independent Republics: US Counter-insurgency Efforts in Colombia, 1959–1965." *Small Wars and Insurgencies* (Winter 1995): 304–327.

———. *The Past as Prologue? A History of US Counterinsurgency Policy in Colombia, 1958–1966*. Carlisle Barracks, PA: Strategic Studies Institute, March 2002.

———. "The Origin of Internal Security in Colombia: Part I-A CIA Special Team Surveys *la violencia*, 1959–1960." *Small Wars and Insurgencies* (Winter 1999): 24–61.

Restrepo, Jorge A., and Michael Spagat. "Colombia's Tipping Point?" *Survival* (Summer 2005): 131–152.

"Retired Generals Fear Consequences of Military Restructuring—Colombian Daily." *BBC Monitoring Americas,* 2 May 2005. http://proquest.umi.com/pqdweb?index=1&did=830857351&SrchMode=2&sid=1&Fmt=3&VInst=PROD&VType=PQD&RQT=309&VName=PQD&TS=1246652143&clientId=5904 (accessed 4 November 2008).

Richani, Nazih. "Caudillos and the Crisis of the Colombian State: Fragmented Sovereignty, the War System and the Privatisation of Counterinsurgency in Colombia." *Third World Quarterly* 28, no. 2 (2007): 403–417.

Richfield, Paul. "Colombia Takes Its First Five Super Tucanos." *C4ISR Journal* (7 December 2006). http://www.c4isrjournal.com/story.php?F=2408477 (accessed 11 March 2009).

Robberson, Tod. "Colombia Plans Offensive." *Salt Lake Tribune,* 4 December 1999. http://proquest.umi.com/pqdweb?index=5&did=46862077&SrchMode=2&sid=1&Fmt=3&VInst=PROD&VType=PQD&RQT=309&VName=PQD&TS=1242675227&clientId=5904 (accessed 5 November 2008).

Robinson, Linda, with Ruth Morris. "Colombia's Messy, Complicated War." *US News and World Report,* 4 September 2000. http://web.ebscohost.com/ehost/detail?vid=7&hid=104&sid=77ac7b0e-8678-4d6b-8bb2-49ddfb6f1153%40sessionmgr109&bdata=JnNpdGU9ZWhvc3QtbGl2ZQ%3d%3d#db=mth&AN=3483857 (accessed 20 May 2009).

Rohter, Larry. "Armed Forces in Colombia Hoping to Get Fighting Fit." *New York Times,* 5 December 1999. http://proquest.umi.com/pqdweb?index=3&did=46865681&SrchMode=2&sid=1&Fmt=3&VInst=PROD&VType=PQD&RQT=309&VName=PQD&TS=1240516114&clientId=5904 (accessed 5 November 2008).

———. "As Colombia Declares an Alert, Rebel Offensive Rages On." *New York Times,* 12 July 1999. http://proquest.umi.com/pqdweb?index=3&did=43068777&SrchMode=2&sid=1&Fmt=3&VInst=PROD&VType=PQD&RQT=309&VName=PQD&TS=1242335255&clientId=5904 (accessed 14 May 2009).

———. "Massacre in Colombia Village Reverberates Paramilitary Terrorism Raises Questions on US Aid for Drug War." *Pittsburg Post—Gazette,* 16 July 2000. http://proquest.umi.com/pqdweb?index=2&did=56460309&SrchMode=2&sid=1&Fmt=3&VInst=PROD&VType=PQD&RQT=309&VName=PQD&TS=1242767033&clientId=5904 (accessed 19 May 2009).

———. "With US Training, Colombia Melds War on Rebels and Drugs." *New York Times,* 29 July 1999. http://proquest.umi.com/pqdweb?index=6&did=43483369&SrchMode=2&sid=4&Fmt=3&VInst=PROD&VType=PQD&RQT=309&VName=PQD&TS=1242413974&clientId=5904 (accessed 5 November 2008).

Saskiewicz, Paul E. "The Revolutionary Armed Forces of Colombia—People's Army (FARC-EP): Marxist-Leninist Insurgency or Criminal Enterprise?" Monterey: CA: Naval Postgraduate School Thesis, December 2005.

Selsky, Andrew. "Colombia Leader Gives his Military More Power." *San Antonio Express–News,* 1 March 2002. http://proquest.umi.com/pqdweb?index=3&did=1171854151&SrchMode=2&sid=3&Fmt=3&VInst=PROD&VType= PQD&RQT=309&VName=PQD&TS=1243540085&clientId=5904 (accessed 28 May 2009).

———. "Colombian Military Commander Resigns." *Washington Post*, 13 November 2003. http://proquest.umi.com/pqdweb?index=7&did=444810391&SrchMode=2&sid=1&Fmt=3&VInst=PROD&VType=PQD&RQT=309&VName=PQD&TS=1246371401&clientId=5904 (accessed 4 November 2008).

Schemo, Diana J. "US Is to Help Army in Colombia Fight Drugs but Skeptics Abound." *New York Times,* 25 October 1997. http://proquest.umi.com/pqdweb?index=23&did=20044995&SrchMode=2&sid=4&Fmt=3&VInst=PROD&VType=PQD&RQT=309&VName=PQD&TS=1239807437&clientId=5904 (accessed 5 November 2008).

———. "Colombia Installs New President Who Plans to Talk to Rebels." *New York Times,* 8 August 1998. http://proquest.umi.com/pqdweb?index=4&did

=32695228&SrchMode=2&sid=19&Fmt=3&VInst=PROD&VType =PQD&RQT=309&VName=PQD&TS=1236966804&clientId=5904 (accessed 5 November 2008).

Skinner, Kristian D. "An Historical Analysis of the Colombian Dilemma." Washington, DC: National Defense University Student Paper, April 2001.

Spencer, David. *Colombia's Paramilitaries: Criminals or Political Force?* Carlisle Barracks, PA: Strategic Studies Institute, December 2001.

———. "Focus—Latin America, Bogotá Continues to Bleed as FARC Find Their Military Feet." *Jane's Intelligence Review,* 1 November 1998. http:// search.janes.com/Search/documentView.do?docId=/content1/janesdata/ mags/jir/history/jir98/jir00815.htm@current&pageSelected=allJane s&keyword=david%20spencer&backPath=http://search.janes.com/ Search&Prod_Name=JIR& (accessed 4 March 2009).

———. "Latin America, A Lesson for Colombia." *Jane's Intelligence Review,* 1 October 1997. http://search.janes.com/Search/documentView.do?doc Id=/content1/janesdata/mags/jir/history/jir97/jir00125.htm@current& pageSelected=allJanes&keyword=david%20spencer&backPath=http:// search.janes.com/Search&Prod_Name=JIR& (accessed 4 March 2009).

———. "Latin America, FARC's Innovative Artillery." *Jane's Intelligence Review,* 1 December 1999. http://search.janes.com/Search/documentView.do? docId=/content1/janesdata/mags/jir/history/jir99/jir00594.htm@current &pageSelected=allJanes&keyword=david%20spencer&backPath= http:// search.janes.com/Search&Prod_Name=JIR& (accessed 4 March 2009).

Sprunk, Darren D. "Transformation in the Developing World: An Analysis of Colombia's Security Transformation." Monterey, CA: Naval Postgraduate School Thesis, September 2004.

"The Warning from Colombia's Generals." *Washington Times,* 30 May 1999. http://proquest.umi.com/pqdweb?index=9&did=41983631&SrchMode =2&sid=3&Fmt=3&VInst=PROD&VType=PQD&RQT=309&VName =PQD&TS=1242327099&clientId=5904 (accessed 14 May 2009).

Torchia, Christopher. "Weakness of Army Prolongs Colombian War." *Colombian,* 2 September 1996. http://proquest.umi.com/pqdweb?index=6&did=153 12706&SrchMode=2&sid=1&Fmt=3&VInst=PROD&VType=PQD& RQT=309&VName=PQD&TS=1241202373&clientId=5904 (accessed 5 November 2008).

Valenzuela, Alfred A., and Victor M. Rosello. "Expanding Roles and Missions in the War on Drugs and Terrorism: El Salvador and Colombia." *Military Review* (March–April 2004): 28–35.

Van Dongen, Rachel. "'Zones' Suspend Colombian Rights; Military Put in Charge of Security, Given Authority Over Civilians." *Washington Times,* 3 December 2002. http://proquest.umi.com/pqdweb?index= 3&did=250 009071&SrchMode=2&sid=3&Fmt=3&VInst=PROD&VType=PQD& RQT=309&VName=PQD&TS=1245781832&clientId=5904 (accessed 4 November 2002).

————. "Colombia's Newest Troops Don't Have to Leave Home, Some 5,000 Troops Eagerly Enlist in a Program that Lets Them Serve in Own Villages." *Christian Science Monitor*, 9 April 2003. http://proquest. umi. com/pqdweb?index=0&did=322434131&SrchMode=2&sid=2&Fmt=3 &VInst=PROD&VType=PQD&RQT=309&VName=PQD&TS=12461 36585&clientId=5904 (accessed 4 November 2008).

————. "Plan Puts Colombia on Offensive; Top US Officials Asked Congress Last Week to Increase the Cap on Troops Allowed in Colombia." *Christian Science Monitor*, 22 June 2004. http://proquest.umi.com/pqdweb?index= 0&did=653680301&SrchMode=2&sid=1&Fmt=3&VInst=PROD&V Type=PQD&RQT=309&VName=PQD&TS=1246456382&clientId =5904 (accessed 4 November 2008).

Vauters, Jason, and Michael L.R. Smith. "A Question of Escalation—From Counternarcotics to Counterterrorism: Analyzing US Strategy in Colombia." *Small Wars and Insurgencies* (June 2006): 163–196.

Vieira, Constanza. "Colombia: Uribe Uses Military to Monitor Opposition Lawmakers." *Global Information Network*, 23 April 2007. http:// proquest.umi.com/pqdweb?did=1259095951&sid=4&Fmt=3&clientId= 5094&RQT=309&VName=PQD (accessed 3 November 2008).

Waghelstein, John D. "Military-to-Military Contacts: Personal Observations— The El Salvador Case." *Low Intensity Conflict & Law Enforcement* (Summer 2001): 1–45.

Waldmann, Peter. "Is There a Culture of Violence in Colombia?" *Terrorism and Political Violence* 19 (2007): 593–609.

Watson, Cynthia. "Civil Military Relations in Colombia: Solving or Delaying Problems?" *Journal of Political and Military Sociology* (Summer 2005): 97–106.

Wilson, Scott. "Colombia Increases Military's Powers; Law Could Threaten US Aid Disbursement." *Washington Post,* 17 August 2001. http://proquest. umi.com/pqdweb?index=15&did=77942545&SrchMode=2&sid=1&F mt=3&VInst=PROD&VType=PQD&RQT=309&VName=PQD&TS= 1243445918&clientId=5904 (accessed 27 May 2009).

————. "Colombia Poised to Install Leader as Rebels Attack; Dozens Dead in Wide-Ranging Offensive on Eve of Hard-Liner's Inauguration." *Washington Post,* 7 August 2002. http://proquest.umi.com/pqdweb ?index=3&did=146818411&SrchMode=2&sid=1&Fmt=3&VInst=PR OD&VType=PQD&RQT=309&VName=PQD&TS=1245852022&c lientId=5904 (accessed 4 November 2008).

————. "Colombia Targeting Rebel Strongholds; More Aggressive US-Backed Strategy Expected to be More Challenging, Brutal." *Washington Post*, 25 January 2004. http://proquest.umi.com/pqdweb?index=4&did=5304 44221&SrchMode=2&sid=1&Fmt=3&VInst=PROD&VType=PQD&R QT=309&VName=PQD&TS=1246389072&clientId=5904 (accessed 4 November 2008).

————. "Colombian Army Ordered into Haven as Rebel Talks End." *Washington Post,* 21 February 2002. http://proquest.umi.com/pqdweb?index=5&did=109434508&SrchMode=2&sid=1&Fmt=3&VInst=PROD&VType=PQD&RQT=309&VName=PQD&TS=1243529560&clientId=5904 (accessed 4 November 2008).

————. "Colombian General Convicted in Killings; Collaboration with Paramilitaries Seen." *Washington Post,* 14 February 2001. http://proquest.umi.com/pqdweb?index=12&did=68578862&SrchMode=2&sid=1&Fmt=3&VInst=PROD&VType=PQD&RQT=309&VName=PQD&TS=1243364549&clientId=5904 (accessed 23 May 2009).

————. "Colombians Ill-Prepared for Prolonged War on Rebels." *Washington Post,* 3 March 2002. http://proquest.umi.com/pqdweb?index=7&did=110002917&SrchMode=2&sid=1&Fmt=3&VInst=PROD&VType=PQD&RQT=309&VName=PQD&TS=1243536257&clientId=5904 (accessed 4 November 2008).

————. "Colombia's Other Army; Growing Paramilitary Force Wields Power with Brutality." *Washington Post,* 12 March 2001. http://proquest.umi.com/pqdweb?index=6&did=69550386&SrchMode=2&sid=2&Fmt=3&VInst=PROD&VType=PQD&RQT=309&VName=PQD&TS=1243365516&clientId=5904 (accessed 4 November 2008).

————. "No Sanctuary from Colombian War; Army Was Absent during Massacre at Village Church." *Washington Post,* 9 May 2002. http://proquest.umi.com/pqdweb?index=5&did=118615149&SrchMode=2&sid=1&Fmt=3&VInst=PROD&VType=PQD&RQT=309&VName=PQD&TS=1243625034&clientId=5904 (accessed 28 May 2009).

————. "US Moves Closer to Colombia's War; Involvement of Special Forces Could Trigger New Wave of Guerrilla Violence." *Washington Post,* 7 February 2003. http://proquest.umi.com/pqdweb?index=3&did=284068341&SrchMode=2&sid=3&Fmt=3&VInst=PROD&VType=PQD&RQT=309&VName=PQD&TS=1246038232&clientId=5904 (accessed 4 November 2008).

————. "War with an Absent Army; In Contested Region, Colombian Government Finds Some Towns Too Dangerous to Protect." *Washington Post,* 3 August 2001. http://proquest.umi.com/pqdweb?index=1&did=76989264&SrchMode=2&sid=3&Fmt=3&VInst=PROD&VType=PQD&RQT=309&VName=PQD&TS=1243367018&clientId=5904 (accessed 4 November 2008).

————. "Which Way in Afghanistan? Ask Colombia for Directions." *Washington Post,* 5 April 2009. http://www.washingtonpost.com/wp-dyn/content/article/2009/04/03/AR2009040302135_pf.html (accessed 6 April 2009).

Zackrison, James L., ed. *Crisis? What Crisis? Security Issues in Colombia.* Washington, DC: Institute for National Strategic Studies, National Defense University, 1999. http://www.ndu.edu/inss/books/books%20-%201999/Crisis%20What%20Crisis%20Eng%20Oct%2099/criscont.html (accessed 11 February 2009).

About the Author

Robert D. Ramsey III retired from the US Army in 1993 after 24 years of service as an Infantry officer that included tours in Vietnam, Korea, and the Sinai. He earned an M.A. in history from Rice University. Mr. Ramsey taught military history for 3 years at the United States Military Academy and 6 years at the US Army Command and General Staff College. Mr. Ramsey is the author of Global War on Terrorism Occasional Paper 18, *Advising Indigenous Forces: American Advisors in Korea, Vietnam, and El Salvador;* Occasional Paper 19, *Advice for Advisors: Suggestions and Observations from Lawrence to the Present;* Occasional Paper 24, *Savage Wars of Peace: Case Studies of Pacification in the Philippines, 1900–1902;* and Occasional Paper 25, *A Masterpiece of Counterguerrilla Warfare: BG J. Franklin Bell in the Philippines, 1901–1902.*